Trees *of* Vancouver

Gerald B. Straley

UBCPress / Vancouver

© UBC Press 1992
All rights reserved
Printed in Canada on acid-free paper ∞

ISBN 0-7748-0406-8

Canadian Cataloguing in Publication Data
Straley, Gerald Bane, 1945-
  Trees of Vancouver

  Includes bibliographical references and index.
  ISBN 0-7748-0406-8

  1. Trees–British Columbia–Vancouver–
Identification. I. Title
QK203.B7S87 1992   582.1609711'33   C92-091006-8

Design: George Vaitkunas

UBC Press
University of British Columbia Press
6344 Memorial Road
Vancouver, BC V6T 1Z2

(604) 822-3259
Fax: (604) 822-6083

*I fondly dedicate this book to my parents,
who taught me to observe*

Contents

Note: Colour plates for trees follow page 104.
Plate numbers for trees found in the
colour plate section are indicated at the end
of the corresponding entry in the text.

Acknowledgments

Thanks are given to many people for help with identifications, locations of specific trees, and other useful comments on this book, as well as hours of thoughtful conversations about our trees. These people include Roy L. Taylor, who was the first person I asked about a tree upon my arrival in Vancouver in 1976, and the following people for 'talking trees' over the years: David Tarrant, Bruce Macdonald, Charles Tubesing, Peter Wharton, Tom Wheeler, John Neill, Roy Forster, David Bar-Zvi, Alleyne Cook, Don Benson, Clive Justice, Doug Justice, Cornelia Oberlander, Brent Robertson, John Worrall, J.C. Raulston, Dan Hinkley, Jim Walker, Russ Archer, Mark Flanagan, and Arthur Lee Jacobson.

For hours of patience while driving around the city and stopping to look at some tree, thanks are due to my late friend Stephen Lipman.

I thank the following people for assistance with identification of specimens of specific groups of trees: George W. Argus (willows), A. Edward Murray, Jr. (maples), Harry Van de Laar (ornamental cherries), and Roger Vick (poplars).

For assistance in compiling data and reading parts of the manuscript, I thank Mary Lou Gillies, Vivian Glyn-Jones, and Mildred Greggor. Sylvia Taylor deserves special thanks for her careful reading of the manuscript and many editorial comments.

I appreciate the pleasant demeanour and co-operation of the staff at UBC Press, Peter Milroy and Jean Wilson in the early stages, and especially Holly Keller-Brohman and George Vaitkunas for the final production and design of the book.

I thank the International Society of Arboriculture for a grant in 1984 to evaluate the commercial potential of trees in the Asian Garden of UBC Botanical Garden.

And, finally, to Katherine Muller, Richard Broder, and Will Beittel, none of whom I have met, but their book *Trees of Santa Barbara* was the original inspiration for me to write this book.

Introduction

Visitors to Vancouver are often impressed by two related qualities of the city. First, it is a very green city and, second, it is a city of trees. The lush verdant colour comes from the grass, which is green throughout most of the year, and from the large number of broad-leaved evergreens and conifers that dominate the landscape. For a few weeks in spring the green is temporarily interrupted by a flush of white to pink when the thousands of flowering cherries, plums, and other ornamental trees lining our streets and in parks and private and public gardens bloom. Greens again cool the city for the warm summer months. Autumn colour is not always good in our wet coastal climate, but many cultivated trees often provide the best colours (better than those of the native deciduous trees) – from the golden yellow of the Tulip Tree and Norway Maple, the red of the Eastern Dogwood, Sourwood and Red Maple, and the purple of the Smoke Tree, to the multicolours of *Cercidiphyllum, Parrotia* and Sweet Gum. The deciduous trees may be relatively dull in winter, but it is then that their branching structure is most prominent and that the subtle differences in bark colours and textures become apparent. In winter the contrasting shades of greens, greys, and golds of the evergreen conifers and broad-leaved evergreens, including English Holly, Cherry Laurel, and Arbutus, dominate our landscape.

A street tree management study conducted by an independent consultant to the Vancouver Parks Board in 1988 estimated that there are over 75,000 street trees in the city. This figure does not include trees on private property or in natural areas, parks, and public gardens. It is a daunting task to keep records on all these trees, much less to maintain them.

Most guides to tree identification concentrate only on the native trees or those cultivated in a wide geographical area. One objective of this book is to include all, or at least a great majority of, native and cultivated trees. It is meant to be used both as a guide for visitors to the city who have a curiosity about our many beautiful trees and by the local people who see the trees every day but may not really 'see' them. However, the primary objective is for the book to be used as a guide by horticulturists, landscape architects, nursery people, naturalists, and, especially, homeowners who want to identify trees or learn how large a certain kind of tree will become after five or ten years or at maturity. One can see if the shrub planted by the front door will become a tree in a few years and quickly outgrow its location. If one wants to see what a 'Kanzan' Japanese Cherry, White Fir or 'Eddie's White Wonder' Dogwood looks like as a specimen tree or street planting, this book will give a few localities where the trees may be seen around the city. A leaf of a maple thought to be a Norway Maple may be compared to the planting along 12th Ave and Burrard St, for instance.

An effort has been made to list all the species and major varieties and cultivars found in the city. Certainly some have been missed and others misidentified, although it is hoped that this number is low. Some very small, or poor, one-of-a-kind specimens have not been included, nor have the very rare trees known to be found only in VanDusen Botanical Garden or the University of British Columbia Botanical Garden. Had all these been included, this book would have doubled in size. Also, many of these were recently planted and are, as yet, quite small. Trees in these two gardens are included only if there is at least one other specimen in a public place or private garden visible from the street. Readers are encouraged to visit these two botanical gardens with their fine, developing collections of trees, most of which are labelled. Trees in private gardens are listed only if they are clearly visible from the street. Therefore, a number of trees in back gardens could have been listed but are not.

The book is divided into two sections, with conifers and related trees (gymnosperms) in the first section and flowering plants (angiosperms) in the second section. Plants are listed alphabetically by family. Within a family, the trees are arranged alphabetically by genus, species, and cultivars. Synonyms and authorities of scientific names and family names have been included for those people who care about this additional information. Common names have been standardized whenever possible. The standard reference for native plants is *Vascular Plants of British Columbia* (Taylor and MacBryde 1977), while that for cultivated plants is *Hortus III* (Bailey Hortorium 1976). In a few cases more than one common name is used, even locally, and both or all of these have been included. A few trees have no common names and the generic name has been used as a common name in these cases. All common names may be found in the index.

Information for each species includes how it differs from similar kinds, a general description, origin, naturally occurring varieties and/or garden-origin cultivars, and the location of one to several specimens in the city. Larger groups, such as pines or oaks, have an introductory paragraph which is included to provide general information about the genus. Drawings of leaves, flowers, fruits, or other distinctive features are included for many of the trees.

The geographical scope of this book includes the City of Vancouver from Boundary Rd west including Pacific Spirit Regional Park, the University Endowment Lands, and the University of British Columbia campus. There are equally interesting trees in all the adjoining municipalities, although none of these are included.

Trees grow, die, are topped, pruned, clipped, limbed-up, moved, run into and over, and are replaced with other trees, or too often, with buildings or parking lots. Do not be surprised if certain ones referred to specifically in this book are no longer there when you look for them. A sad aspect of compiling this book has been to see trees lost in the city during my years here. The many notable losses include: a large and well-known *Paulownia* on Point Grey Rd, removed by the owner; a pair of Japanese Pagoda Trees on 12th Ave at Granville St and old Black Walnuts on 12th Ave at Cambie St, removed for street widening; the largest *Acer cissifolium* in the city, a magnificent old specimen at Vancouver General Hospital, cut down for hospital expansion; the only *Pinus sabiniana* (Digger Pine) in the city, on the corner of 2nd Ave and Larch St, removed in 1991; and several trees in the Old Arboretum on the UBC campus, recently removed for building construction.

The definition of a tree is not always clear-cut. There is no problem saying that big trees with single trunks are definitely trees, while small, bushy plants with several trunks and limbs down to the ground are shrubs, but where does one stop and the other start? Included here are some plants such as lilacs and rhododendrons that are usually thought of as shrubs although they can definitely become tree-like with age. Others such as laurels (*Prunus laurocerasus*) and cedars (*Chamaecyparis lawsoniana*) used as hedges remain as such only because they are pruned. They will definitely become trees in time. Some shrubs can be encouraged to become tree-like more quickly by pruning off the lower branches and allowing only one or a few trunks to grow. So, a tree is not always a definite entity, but rather a combination of genetic make-up, size, form, and age.

There are unexpected difficulties in identifying even well-known trees, as they may take on a very different character when grown away from their natural habitat. This is especially true for some of the elms, pines, and oaks. Some closely related trees found in different parts of the world are easily distinguished when one knows their place of origin, but in a garden situation they may look so much alike that it is difficult to separate them. Notable among species pairs in our gardens that are very similar are English Yew (*Taxus baccata*) and Western Yew (*Taxus brevifolia*), Cedar-of-Lebanon (*Cedrus libani*) and Atlas Cedar (*Cedrus atlantica*), and our native Vine Maple (*Acer circinatum*) and the Japanese Fullmoon Maple (*Acer japonicum*).

Trees that are difficult to identify include the cultivars of some species of trees, notably Japanese flowering cherries, crabapples, and Lawson cypresses. There are dozens, if not hundreds of different cultivars of these species, many known only from vague descriptions in references. They were often planted for only a short time and are no longer available commercially. It becomes a nearly impossible task to identify these cultivars.

Another complication in identifying trees is deciding on what is a typical leaf. Leaves differ from population to population, from tree to tree, or even from branch to branch. Sun leaves, or those toward the tops of trees and the ends of branches, tend to be smaller, thicker and often less-lobed than shade leaves, those growing in shade inside the tree and closer to the main limbs and trunk. Leaves growing from vigorous shoots, known as water sprouts, or from trunk sprouts are often much larger and often very different from those on upper limbs.

A number of trees commonly planted elsewhere in North America or other temperate parts of the world are surprisingly absent from Vancouver, or are known from only an individual or two in UBC Botanical Garden or VanDusen Botanical Garden. Examples of notably absent trees in Vancouver are White Oak (*Quercus alba*), Chinese Elm (*Ulmus parvifolia*), Pecan (*Carya illinoinensis*), Balsam Fir (*Abies balsamea*), Red Pine (*Pinus resinosa*), and Lace-bark Pine (*Pinus bungeana*).

Many old trees are included, for which there are interesting stories (who planted them, when or why), but these are not specifically mentioned. This is not meant to be a heritage tree book. Information about the heritage trees can be found by consulting an unpublished manuscript entitled 'Vancouver's Heritage Tree Inventory – A Collection of Great Trees in the City' (1983) by Elizabeth Whitelaw and Clarence Sihoe, a copy of which is kept in the VanDusen Botanical Garden Library.

Where to See Trees in Vancouver

Obviously trees are everywhere in the city, but there are a few specific areas where a wide variety may be observed. The Old Arboretum on the University of British Columbia campus has a nice collection of very large specimens and a number of unusual trees found nowhere else in the city. The planting of the Arboretum began with native trees in 1916 and exotic trees after the Second World War, but unfortunately many of the trees have been lost to parking lots and building construction. The Arboretum was originally part of the Botanical Garden, when the Garden occupied most of the campus. The remains of the Arboretum are just west of West Mall, from the Fraser River Parkade south to University Blvd, behind the Ponderosa Cafeteria. In addition, there are many other interesting trees scattered around the campus. A map listing the trees in the Old Arboretum follows on page xviii and can be used for a self-guided walk.

The UBC Botanical Garden has a huge collection of trees, although many of them are still small since the plantings began only in 1972. The David C. Lam Asian Garden, located between old and new SW Marine Drives, is the richest area in the garden for unusual trees, especially maples, magnolias, mountain ashes, and rhododendrons. Virtually all of the trees native to British Columbia may be seen somewhere in the Native Garden, and other unusual trees may be found in other garden areas.

The VanDusen Botanical Garden at Oak St and 37th Ave is likewise the home of a large collection of interesting and unusual plants in beautiful garden settings. Most of the trees on the site have been planted since 1972 and are still quite young. However, they give one a sense of how quickly small trees grow. Most of the trees are labelled at both UBC and VanDusen Botanical Gardens.

Stanley Park is an excellent place to see a number of both common and rare trees, especially in the areas around the Zoo, Parks Board Office, and the Pitch & Putt Golf Course. Queen Elizabeth Park, with its Arboretum and Quarry Garden, also has a fine collection of trees, many found nowhere else in the city.

The Crescent in Shaughnessy is a large circle, the centre of which is planted with many old and unusual trees. There do not seem to be any definite records of who planted the trees or when, but it is a good place to see how large some of these trees become in 50 to 100 years. There is a map of the trees in The Crescent on page xxii. There are also some trees, such as Yellowwood (*Cladrastis*), found there and nowhere else in the city.

There are many good city street plantings, but the best are found on streets with wide medians. King Edward Ave, West 16th Ave, and Cambie St are among the most notable for a diversity of interesting trees.

For other localities in the city, obtain a city street map and a UBC campus map to accompany this book. Take different routes home from work or drive around on the weekend.

Happy hunting!

Trees in the Old Arboretum at UBC

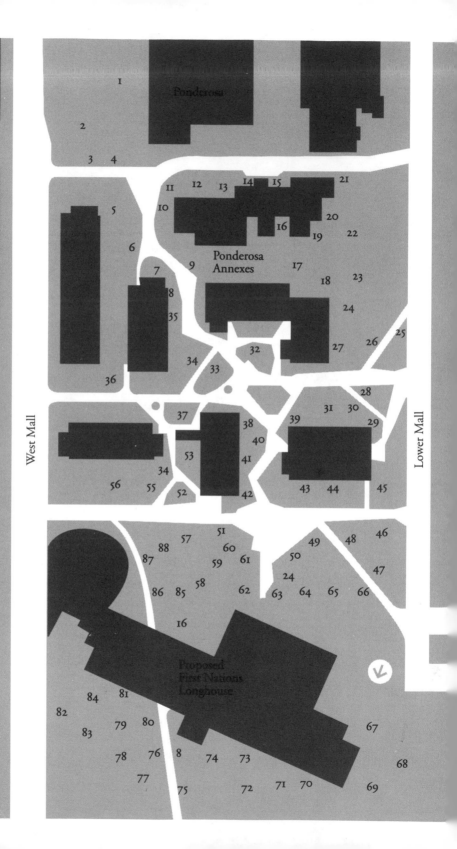

1 *Pinus ponderosa* – Ponderosa Pine
2 *Chamaecyparis lawsoniana* – Lawson Cypress
3 *Pinus coulteri* – Coulter Pine
4 *Parrotia persica* – Persian Ironwood
5 *Oemlaria cerasiformis* – Indian Plum*
6 *Acer circinatum* – Vine Maple
7 *Populus balsamifera* ssp. *trichocarpa* – Black Cottonwood
8 *Pinus sylvestris* – Scots Pine
9 *Populus tremuloides* – Quaking Aspen
10 *Maclura pomifera* – Osage Orange
11 *Chamaecyparis nootkatensis* – Yellow Cedar
12 *Thuja plicata* – Western Red Cedar
13 *Abies grandis* – Grand Fir
14 *Pseudotsuga menziesii* – Douglas Fir
15 *Picea sitchensis* – Sitka Spruce
16 *Betula papyrifera* – Paper Birch
17 *Fraxinus excelsior* – Common Ash
18 *Rhamnus purshiana* – Cascara
19 *Alnus rubra* – Red Alder
20 *Tsuga mertensiana* – Mountain Hemlock
21 *Tsuga heterophylla* – Western Hemlock
22 *Alnus crispa* ssp. *sinuata* – Sitka Alder
23 *Styrax obassia* – Fragrant Snowbell
24 *Celtis occidentalis* – Hackberry
25 *Prunus × yedoensis* – Yoshino Cherry
26 *Diospyros virginiana* – Common Persimmon
27 *Fraxinus pennsylvanica* var. *lanceolata* – Green Ash*
28 *Acer glabrum* – Douglas Maple
29 *Halesia tetraptera* – Silver-Bell Tree
30 *Salix matsudana* 'Tortuosa' – Corkscrew Willow
31 *Cornus nuttallii* – Western Dogwood
32 *Cornus florida* – Eastern Dogwood
33 *Tilia americana* – American Linden or Basswood
34 *Pterocarya stenoptera* – Chinese Wingnut
35 *Amelanchier laevis* – Service-Berry
36 *Quercus petraea* – Sessile Oak
37 *Fraxinus nigra* – Black Ash*
38 *Fraxinus ornus* – Flowering Ash
39 *Fraxinus americana* – White Ash
40 *Cornus sanguinia* – Red-Branched Dogwood*
41 *Tilia petiolaris* – Pendent Silver Linden
42 *Acer saccharum* – Sugar Maple
43 *Euonymus alata* – Winged Spindle Tree*
44 *Ilex aquifolium* – English Holly
45 *Ailanthus altissima* – Tree-of-Heaven
46 *Robinia pseudoacacia* – Black Locust
47 *Quercus macrocarpa* – Bur Oak

* indicates rare trees included here but not discussed in the text.

Trees in The Crescent, Vancouver

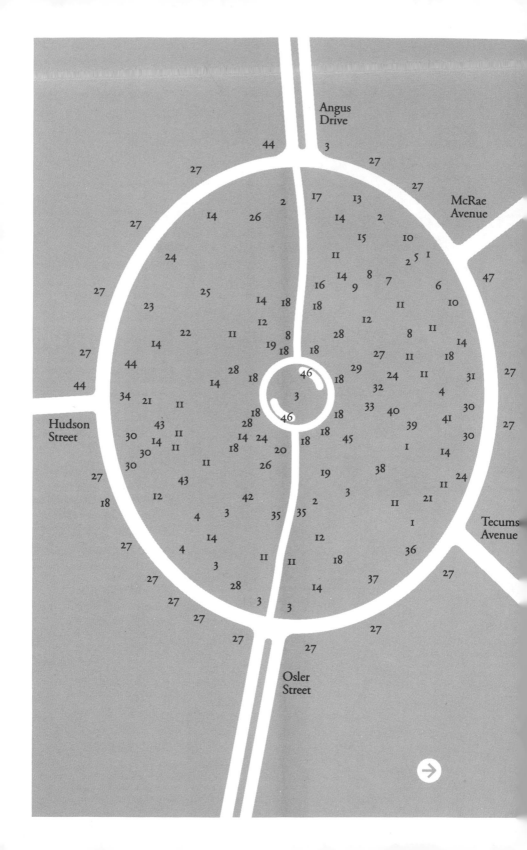

1   *Styrax japonicus* – Japanese Snowbell
2   *Fraxinus ornus* – Flowering Ash
3   *Acer saccharum* – Sugar Maple
4   *Aesculus × carnea* – Red Horse-Chestnut
5   *Taxus baccata* – English Yew
6   *Taxus baccata* (cultivar) – Spreading English Yew
7   *Fraxinus excelsior* – Common Ash
8   *Euonymus alata* – Winged Euonymus*
9   *Chionanthus virginica* – Fringe-Tree
10   *Picea pungens* – Blue Spruce
11   *Ulmus carpinifolia* – Smooth-Leaved Elm
12   *Picea abies* – Norway Spruce
13   *Malus floribunda* – Japanese Crab Apple
14   *Ilex aquifolium* – English Holly
15   *Acer palmatum 'Atropurpureum'* – Japanese Maple
16   *Crataegus laevigata 'Plena'* – Double English Hawthorn
17   *Sequoia sempervirens* – Coast Redwood
18   *Tilia platyphyllos* – Large-Leaved Linden
19   *Chamaecyparis pisifera* – Sawara Cypress
20   *Rhododendron ponticum* – Ponticum Rhododendron
21   *Acer saccharinum* – Silver Maple
22   *Prunus cerasifera 'Atropurpurea'* – Purple-Leaved Plum
23   *Acer circinatum* – Vine Maple
24   *Chamaecyparis lawsoniana 'Stewartii'* – Stewart Golden Cypress
25   *Fagus sylvatica 'Atropunicea'* – Copper Beech
26   *Tsuga canadensis* – Canadian Hemlock
27   *Aesculus hippocastanum* – Common Horse-Chestnut
28   *Oxydendrum arboreum* – Sourwood
29   *Acer pseudoplatanus* – Sycamore Maple
30   *Chamaecyparis lawsoniana 'Allumii'* – Pyramidal Blue Lawson Cypress
31   *Acer pseudoplatanus 'Atropurpureum'* – Purple Sycamore Maple
32   *Euonymus europaeus* – European Spindle-Tree
33   *Pseudotsuga menziesii* – Douglas Fir
34   *Picea orientalis* – Oriental Spruce
35   *Buxus sempervirens* – Common Box*
36   *Prunus serrulata 'Takasago'* – Takasago Japanese Cherry
37   *Chamaecyparis lawsoniana 'Fletcheri'* – Fletcheri Lawson Cypress*
38   *Syringa vulgaris* – Common Lilac (white)
39   *Juniperus virginiana* – Red Cedar
40   *Ailanthus altissima* – Tree-of-Heaven
41   *Fraxinus pennsylvanica* – Red Ash
42   *Fagus sylvatica* – European Beech
43   *Cladrastis kentukea* – Yellowwood
44   *Acer platanoides* – Norway Maple
45   *Rhododendron maximum* – Rosebay*
46   *Cornus 'Eddie's White Wonder'* – Eddie's White Wonder Dogwood
47   *Prunus × yedoensis* – Yoshino Cherry

---

* indicates rare trees included here but not discussed in the text.

Gymnosperms:
Cone-Bearing and Related Trees

Araucariaceae – Araucaria Family

Araucaria araucana (Mol.) Koch
Monkey-Puzzle Tree

This large evergreen tree native to Chile and Argentina is the only member of a largely subtropical family of graceful, symmetrical trees that is hardy in the Vancouver area. The Monkey-Puzzle Tree is closely related to the Norfolk Island Pine (*Araucaria heterophylla*) that is seen in the tropics or is cultivated indoors in our area as a pot plant. Monkey-Puzzle is characterized, especially when young, by the perfect symmetry of its whorled branches and the very artificial, almost plastic, look and feel of the wedge-shaped leaves. The leaves are very hard and sharply pointed and are not pleasing to the touch. Leaves are produced all around the rounded limbs and are retained on the branches for many years. Older trees lose some of the tiered symmetry, taking on a less formal but more pleasing shape. Some individuals tend to loose their branches from the ground up as they mature, while others may retain them to near ground level. The trees are either male or female. Male cones are about 15 cm long, have slender upwardly curved scales, and are borne in large quantities. Female trees, which are not common in Vancouver gardens, bear much larger, round cones that are nearly as large as a basketball. These break apart at maturity, releasing the seeds.

The Monkey-Puzzle was especially popular in the 1920s and 1930s, so there are a number of large specimens around the city. The largest and most spectacular male tree is on 33rd Ave just east of Granville St; there are also large males on Blanca St at the corner of 5th Ave, at King Edward Ave and Dunbar St, at Alexandra St and 26th Ave, and a good male with limbs nearly to the ground on the south side of 11th Ave between Blanca St and Tolmie St. The largest female is on the north side of Pandora St just west of Kaslo St; there are females near each other in Kits Point (one on Ogden Ave east of Maple St and one on McNicholl Ave at Arbutus St), and on the SE corner of 2nd Ave and Tolmie St. •1

Cupressaceae – Cypress Family

Calocedrus decurrens (Torr.) Florin
(*Libocedrus decurrens* Torr.)
Incense Cedar

This beautiful evergreen is native from coastal Oregon to northern Baja California east to the Sierra Nevada of California and western Nevada. It has a distinctly columnar habit and drooping, flattened sprays of slender, bright green branches. The fragrance emitted by the crushed branches is very pleasant. The small, yellow male cones are borne in such profusion as to be noticeable from a distance in late winter (usually in February or March). The female cones are of a distinctive shape, made up of 2–3 cm long flattened scales.

It is occasionally cultivated here. There are large specimens on the NW corner of Marguerite St and 32nd Ave, in the median strip of East Mall just north of University Blvd at UBC, and on the north side of 54th Ave just west of Adera St. There are also nice specimens in the centre of the block along Commercial Dr between Charles St and William St, in front of Buchanan Bldg on the north end of Main Mall at UBC, at the corner of Beverly Cres and 32nd Ave, on the SE corner of 5th Ave and Alma St, and three at the Canadian National Institute for the Blind Bldg on 37th Ave and Sophia St.

Chamaecyparis – False Cypresses

In Vancouver these trees, especially the Lawson Cypress, are commonly called cedars. They are seen in a myriad of forms and are used as specimen trees and, often, as hedges. The leaves

Twigs and cones
of *Chamaecyparis*

1 *C. lawsoniana*
2 *C. nootkatensis*
3 *C. obtusa*
4 *C. pisifera*

The drawings of the cones are 3 times the size of the branches.

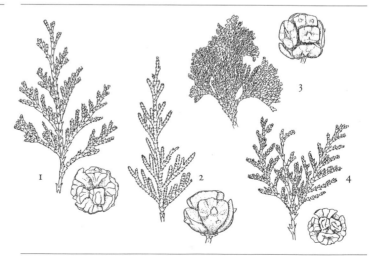

are mostly scale-like, but young trees of some species and cultivars may have longer, curved, needle-like leaves. The small cones are often produced abundantly, are up to about 1 cm in diameter, and are made up of umbrella-shaped (peltate) scales. These scales will easily separate the false cypresses from *Thuja* species, which have flat cone scales.

Chamaecyparis lawsoniana (A. Murr.) Parl.
Lawson Cypress or Port Orford Cedar

This evergreen tree grows as tall as 70 m in nature and occurs in a very narrow belt along the Pacific Coast from southwestern Oregon to northwestern California. It is harvested for timber in its native area and is now much more common in gardens than in the wild. A great number of different cultivated forms have been selected and used in the landscape as specimen shrubs, trees, or, most often, for tall hedges. It is likely our most commonly cultivated evergreen. In cultivation, even old, large trees tend to keep their branches nearly to the ground. Most trees bear the rounded bluish brown or purplish brown cones in great profusion. There are numerous cultivars and unnamed seedlings which show variable growth habits and colours. The most common form found in the wild is dull green, but individual trees vary greatly, even in nature. Locally, the trees suffer badly from a root-rot caused by the fungus *Phytophthera lateralis,* causing the trees to turn brown, usually fairly quickly, and die. There is no effective control, and we may have fewer of these trees around if the disease continues to spread.

Among the green wild-type trees are those on the north side of The Crescent (nearest Tecumseh Ave), one on the sw corner of Balfour Ave and Cartier St (with the greyer 'Allumii' for comparison), and several in Almond Park (east of Alma St at 11th Ave).

There are a number of distinctive cultivars grown in the city, including the following:

'Allumii,' Pyramidal Blue Lawson Cypress – The most common blue-grey cultivar in Vancouver. As a young tree it is an inverted-cone shape with stiff, upright, bright blue-grey foliage, but it becomes duller and of a very different habit as it matures, although even old trees may keep some of the juvenile growth toward the base of the tree. Old specimens are of a broad columnar form with horizontal branches drooping just at the tips. It is similar to 'Intertexta,' but 'Allumii' is usually a narrower tree with bluer foliage. Among the many in the city are three on the north rim of the quarry at Queen Elizabeth Park (with a golden form, 'Stewartii'), one on the sw corner of

Balfour Ave and Cartier St (with a typical wild green form, for comparison), a street planting on both sides of 14th Ave between Cypress St and Maple St, several in The Crescent, two large ones on the NE corner of 39th Ave and Laburnum St, and one on the west side of Laburnum St between 39th Ave and 41st Ave. •2

'Erecta' – The largest of the commonly grown cultivars, with a very broad, dense pyramid shape and soft, bright green foliage. It is not often planted today, but there are many massive old specimens around Vancouver, including a long row of large trees on the north side of 16th Ave from Arbutus St to Cypress St, several on Selkirk St between Matthews Ave and Balfour Ave, and several at the SW corner of the Chemistry Bldg on Main Mall on the UBC campus. •2, 70

'Intertexta' – A popular old cultivar that originated in Scotland in 1869. Widely grown here, it is intermediate in colour between the green wild forms and the blue forms such as 'Allumii,' and forms a slightly broader columnar tree than 'Allumii' at maturity. Old specimens look as if they have been dropped so that the bottoms of the tree spread out wider than the limbs just above the base. It, too, may retain some more erect juvenile limbs near the base of the tree. Locally, the many plantings include dozens in the median of King Edward Ave from Cambie St west to Arbutus St and in the median of 16th Ave from Trafalgar St to Crown St, and a large grove on the east side of Winona Park along Columbia St between 59th Ave and 62nd Ave.

'Pendula,' Weeping Cypress – This cultivar looks so much like the typical form of our native Western Yellow Cedar (*Chamaecyparis nootkatensis*) from a distance that the two are not easily separable unless examined closely. The bright green foliage hangs in long pendulous sheets. Specimens of this cultivar around the city include large ones at Point Grey Secondary School on 37th Ave and East Blvd, on 19th Ave at Highbury St, and south of the Quarry Garden in Queen Elizabeth Park. •3

'Stewartii,' Stewart Golden Cypress – An upright, tall, golden-coloured form that is commonly raised here. There are many old specimens throughout the city, including a nice row on both sides of 15th Ave from Cypress St to Arbutus St, and several specimens in The Crescent in Shaughnessy, in Stanley Park north of the Rose Garden, and on the south side of 4th Ave just east of Blanca St. •4

'Wissellii' – A very columnar form with open, sparse growth, fern-like branchlets, and dark, grey-green foliage. It is one of the most distinctive of locally grown cultivars. The male cones are reddish and borne abundantly in March to April, often giving the trees a reddish cast at a distance. There are nice specimens of this cultivar on the north side of 19th Ave between Wallace St and Highbury St, the west side of Granville St between Balfour St and Laurier Ave, large ones on the north side of 49th Ave between Laburnum St and Cypress St, and on the SE corner of Kitsilano Recreational Centre at 12th Ave and Larch St.

There are a number of other cultivars that are possibly grown in the city, including 'Lutea,' 'Argentea,' 'Coerulea,' and 'Glauca.'

Chamaecyparis nootkatensis (D. Don) Spach
Yellow Cedar or Nootka Cypress

This large tree with pendulous branchlets and a distinct yellow-green colour grows naturally from Alaska to Oregon. It does not grow naturally down to sea level, but it can be seen in the wild around the ski areas on any of the local mountains. The pendulous habit, with generally more rounded branchlets and cones with fewer scales, separates it from the other *Chamaecyparis* species and cultivars seen locally in gardens.

This tree is common here, although less so than the other species in the genus, which is probably a reflection of its being native and therefore generally considered less desirable. At UBC, there is one fine, old, very typical specimen in the Old Arboretum, a slightly smaller one on the NW corner of the Physics Bldg (south of the Main Library), and several small ones, including two between MacMillan Bldg and the Barn Snack Bar and on the south side of the Anthropology and Sociology Bldg just north of NW Marine Dr. There is a large twin-trunked tree on the north side of 45th Ave at Balsam St, and there are some small trees in the Pacific Northwest Section of VanDusen Botanical Garden.

'Pendula,' Weeping Yellow Cedar – This form is even more pendulous than the typical wild form, has more irregular, sparse growth, especially when young, and has been planted more often in recent years than the wild forms. There are two moderately large trees with the characteristic drooping branches in McCleery Park at the corner of Marine Cres and 49th Ave, three young trees along the north side of Point Grey Rd and Cornwall Ave (one east of Collingwood St, one east of Bayswater St, and a third at Stephens St), and two young trees just east of the Physick Garden in UBC Botanical Garden.

Chamaecyparis obtusa (Siebold & Zucc.) Endl.
Hinoki Cypress

This Japanese evergreen tree, which grows to 40 m tall, is usually seen in cultivation as a small graceful shrub. There are also a number of dwarf and golden varieties. The foliage is bright green and is produced on thick, but soft, fan-like branchlets. The cones are a dark, rich brown. Older specimens exhibit attractive reddish brown bark in narrow vertical strips, but this is not usually seen on our local, mostly young, specimens.

Large specimens are surprisingly rare here. Probably the largest one in the city is in a garden on the west side of Collingwood St near the lane between 23rd Ave and 24th Ave. Some of the many small specimens that are beginning to become tree forms are located sw of the Main Library on the UBC campus; there are also two tall ones in a garden on the SE corner of Cambie St and 35th Ave, one in front of the church on the sw corner of 10th Ave and Trutch St, one on the south side of Point Grey Rd between Alma St and Highbury St, and one on the sw side of Burnaby St between Bute St and Jervis St.

'Crippsii,' Cripps Golden Cypress – This is the relatively common golden form, usually showing brighter yellow toward the tips of branches and greener inside. There are specimens on the north side of King Edward Ave just east of Valley Dr, two on the NE corner of 5th Ave and Arbutus St, one on the south side of 2nd Ave between Arbutus St and Maple St (with a *Thuja plicata* 'Zebrina' for comparison), and one by the entrance to the Pitch & Putt Golf Course in Queen Elizabeth Park.

Chamaecyparis pisifera (Siebold & Zucc.) Endl.
Sawara Cypress

A Japanese evergreen very commonly planted in many different forms in our gardens. The most common one is rather columnar in outline with somewhat yellow-green foliage. The best characteristic for distinguishing this tree from other similar evergreens is that it is slightly more prickly to the touch because the tips of the scale-like leaves point out from the branches, especially on the underside. Small, rounded, coppery brown cones are borne in profusion on many trees.

They are usually seen here as hedge plants or as foundation plantings, although the trees soon outgrow their location. The typical wild form with green foliage is often called 'Plumosa.' There are several large trees on 52nd Ave between Fraser St and St. George St, in the middle of The Crescent near Hudson St in

Shaughnessy, around the monument in Memorial South Park off 41st Ave, along Cornwall Ave at Kitsilano Park, and on the NE corner of 2nd Ave and Trutch St.

'Filifera,' Thread Cypress – This is the adult foliage form with long, rounded twigs composed of scale-like leaves. The slender twigs are usually of variable lengths and droop from the larger branches in loose fan-shaped clusters. It is usually a shrubby form, but may become tree-like as seen in several trees NW of the Main Library at UBC, several at the Aberthau Cultural Centre at 2nd Ave and Trimble St, and on the SE corner of Nanton Ave and Pine Cres.

'Filifera Aurea,' Golden Thread Cypress – This is the golden-tipped form of 'Filifera.' There is a large tree on the north side of 32nd Ave between Hudson St and Selkirk St, and one at the Aberthau Cultural Centre at 2nd Ave and Trimble St.

'Plumosa Albopicta' – This is not a very attractive form at a distance because it looks sickly, but the varied tufts of pure white, mixed white and green, and mostly pure green foliage are attractive at close range. It is not a very common cultivar here, but there is a large specimen behind a golden form of Lawson Cypress on the east side of Hudson St between 43rd Ave and 45th Ave, a large one on the SE corner of 33rd Ave and Blenheim St, and a row on the north side of Memorial Rd west of Main Mall at UBC.

'Plumosa Aurea,' Golden Plume Cypress – This is the golden-tipped form similar to the wild green ones. There are several on the grounds of Point Grey Secondary School on East Blvd at 37th Ave, on the NW corner of 57th Ave and Marguerite St, and at the Aberthau Cultural Centre at 2nd Ave and Trimble St.

Var. *squarrosa* Beissn. & Hochst., Moss Cypress – This naturally occurring silvery form of the Sawara Cypress looks very different from the typical green form. The trees quickly become very large and are attractive as a specimen tree. The leaves are longer and silvery grey, forming soft fan-like sprays. Unlike the typical green form, it rarely, if ever, produces cones. It was very popular a few decades ago and, as a result, there are many old specimens around the city. There are large trees in Point Grey Park on Point Grey Rd near Macdonald St, on the north side of 10th Ave between Fir St and Pine St, on Main St at 36th Ave (at the Canadian National Institute for the Blind Bldg), on the north side of 31st Ave at Puget Dr, on Matthews St between Alexandra St and Marguerite St, and a very large individual on 20th Ave between Highbury St and Dunbar St.

×*Cupressocyparis leylandii* (A.B. Jacks. & Dallim.) Dallim. & A.B. Jacks.
Leyland Cypress

This artificial intergeneric hybrid between the Western Yellow Cedar (*Chamaecyparis nootkatensis*) and Monterey Cypress (*Cupressus macrocarpa*) originated in England in 1888. It is the fastest growing cone-bearing tree in the world. A number of different clones have been selected and it is now often planted for reforestation throughout warm-temperate parts of the world. It is more like the Monterey Cypress, but is hardier, and its irregular growth pattern, with distinctive stiff, outwardly point-ing branches, gives it a very ungainly look. The small branchlets are slightly flattened. Cones are intermediate between the two parents, but most of the specimens in Vancouver are as yet too young to produce cones. It is some-times planted as a hedge but looks good only if a very tall hedge is needed and if it is kept clipped when young.

Most of the specimens in the city are young. There are several along South Mall just north of Agronomy Road at UBC, four young (but rapidly becoming large) trees on the north side of Creelman Ave at Walnut St in Kitsilano Point, two on the north side of 12th Ave west of Oak St, a long row along the lane on the north side of VanDusen Botanical Garden from Oak St westward, as well as along the western end of the garden along 33rd Ave, a fairly large specimen on the south side of 30th Ave between Crown St and Wallace St, and one on the north side of 3rd Ave in the middle of the block between Trafalgar St and Larch St.

Cupressus macrocarpa Hartweg
Monterey Cypress

This conifer is native to a very small area of coastal California, the Monterey Peninsula, but is now widely planted in other parts of the United States, Australia and New Zealand, South America, and southern Europe as a fast growing windbreak and for reforestation. In the wild, it has a symmetrical pyramidal shape when young but becomes a flat-topped, asymmetrical tree growing to 25 m tall at maturity. A lone tree on a rocky point overlooking the Monterey Peninsula certainly ranks as one of the most photographed individual trees in the world. The small scale-like leaves are bright green and closely resemble those of a *Chamaecyparis* or *Juniperus*. The branchlets are rounded and tend to spread out in all directions from the branches, in contrast to the more flattened sprays of the branchlets of the

Western Yellow Cedar. The 2–4 cm cones are much larger than those of the Western Yellow Cedar and are made up of 8–12 hard woody scales.

It is sometimes planted here but usually does not survive more than a few years. We are just a bit too far north and most trees are killed during our coldest winters. The trees grow much better even as close as the Victoria area. There is one relatively large specimen in a yard on the NE corner of 49th Ave and Vine St. It must be a hardy individual as it was started from seed collected in San Francisco in 1971 and seems to have suffered little damage in any recent cold winters.

Juniperus – Junipers

A very common part of vegetation in the Northern Hemisphere, the junipers range from very low shrubs to quite large trees. Some species are commonly called cedars, especially the eastern North American *Juniperus virginiana,* which is the common source of the fragrant wood for cedar chests. Junipers have two types of leaves: longer, needle-like leaves and short, overlapping, scale-like leaves. Some species have the needle-like leaves only when they are young, producing scale-like leaves as they mature. Other species may have only one kind throughout their life or a mixture of both types of leaves. The small cones of junipers contain one to several seeds and remain slightly fleshy and berry-like at maturity, never becoming dry and papery or woody as do most other cones. The cones have a characteristic aroma when crushed and some are used for flavourings in cooking or in alcoholic beverages, especially gin.

Many selections have been made from the seventy or so species and brought into cultivation, and these are commonly seen in landscapes in temperate parts of the world. The larger tree forms often show a preference for drier climates, so are not common in our local gardens.

Juniperus chinensis L.
Chinese Juniper

This evergreen is native to China, Mongolia, and Japan and has been in cultivation for centuries. The leaves are variable, with all short needle-like leaves, all overlapping scale-like leaves, or a combination of both on the same plant. In particular, younger plants have more of the needle-like leaves. There are many cultivars of various sizes, shapes, and colours, many of which are important in our landscapes. In nature, the plants may become trees to more than 20 m tall. The berries are 5–10 mm across and covered with a glaucous bloom. Tree like forms are not common in cultivation in Vancouver, with most of those cultivated being low, spreading forms.

'Keteleeri' – This is the only cultivar in the city that generally becomes tree-like. It has attractive, dark green, irregular growth and large blue-grey fruits borne in profusion. It becomes a relatively large tree but keeps its limbs to the ground until quite old.

There is a row of relatively large ones NE of the Rose Garden in Queen Elizabeth Park, several by the Sociology Bldg at UBC, two near the entrance to Totem Park Residences at UBC, and a tall slender one in Stanley Park on the eastern wall of the Vancouver Aquarium by the administrative offices.

Juniperus squamata D. Don 'Meyeri'
Meyer Juniper

The cultivar 'Meyeri' is the only form of this shrub or tree commonly cultivated locally. It was introduced from a garden in China by Frank Meyer, a well-known American plant collector. This cultivar is normally a small, multi-trunked tree with all needle-like leaves. Other forms may become trees up to 25 m tall. The needles are dark green beneath and silvery above. Some forms produce fleshy reddish brown cones that are single seeded, but 'Meyeri' does not usually produce cones, at least locally.

Among the larger tree-like specimens in the city are individuals on the south side of 20th Ave between MacKenzie St and Carnarvon St, on the SE corner of 16th Ave and Blenheim St, on the east side of Main St between 61st Ave and 62nd Ave, on the NW corner of 54th Ave and Heather St, on the north side of 40th Ave between Blenheim St and Collingwood St, and several in the Sino-Himalayan Garden at the VanDusen Botanical Garden.

Juniperus virginiana L.
Red Cedar

This is a good example of one of the many different plants that is known as cedar. It is a common eastern North American evergreen growing to 30 m or more. The needles are very similar to those of the Chinese Juniper but the oval blue-grey berries are about half the size. There are a number of cultivated forms with needle-like, or scale-like, or a combination of both types of leaves, and with variable growth forms and needle colours.

It is rare here. There is a large, grey-needled, tree form with a very nice trunk on the north side of The Crescent toward Tecumseh Ave.

Platycladus orientalis (L.) Franco (*Thuja orientalis* L., *Biota orientalis* [L.] Endl.)
Oriental Arborvitae or White Cedar

This bushy, evergreen shrub or small tree, native to China and Korea, closely resembles the American Arborvitae (*Thuja occidentalis*) in its growth habit, size, colour of foliage, and in its use in gardens. However, the branches are held upright in vertical sprays, rather than horizontally; the cones have a distinctive thickened, hooked, horn-like ridge on each cone scale; and the seeds are relatively large, hard, and wingless. These characteristics separate it from *Thuja,* the genus in which it was originally placed. The compact forms (usually called Globe Arborvitae) are often used in foundation plantings, which they quickly outgrow. It is tolerant of a wide range of growing conditions, including tropical climates.

Large tree forms are not very common locally. The larger ones include specimens on Osler St near The Crescent, on Angus Dr between The Crescent and Granville St, on the east side of the footbridge between the causeway and the yacht club in Stanley Park, and large ones in the Old Arboretum and along West Mall just west of the Graduate Student Centre at UBC.

Thuja occidentalis L.
American Arborvitae

This small tree, growing to 20 m tall and occurring from Nova Scotia to North Carolina and Illinois, cannot match the massiveness of its western relative, the Western Red Cedar (*Thuja plicata*). It is often cultivated in our gardens, usually in one of its many dwarf, columnar, or pyramidal forms. The branchlets tend to flare out horizontally in flattened fan-like groups. Its most distinctive characteristic is the dull yellow-green colour in summer and the very dull yellow-brown colour in winter. It looks as if it is near death in winter. The cones have flattened scales and are smaller than those of the Western Red Cedar, and glands on the underside of the scale-like leaves are more prominent. Most of the specimens of the wild form around town are relatively small.

It is not very commonly planted outside parks and gardens, and most of these are small specimens. The larger ones include

specimens in Devonshire Park on Devonshire Cres between
Hudson St and Selkirk St, on Wallace St between 18th Ave and
19th Ave, on Hosmer Ave between Cypress St and Pine St, in
the median of Angus Dr between Hosmer Ave and Granville St,
on the north side of 3rd Ave between Macdonald St and Balsam
St at Tatlow Park, and just SW of the Main Library on the
UBC campus.

'Fastigiata' ('Columnaris') – A very common cultivar that
forms tall green columns, usually with a single trunk and short
side branches. It is seen throughout the city. Although it is
usually shrub-like with limbs to the ground, older specimens
develop a distinct trunk, but remain very narrow. Among the
larger ones around the city are individuals on the south side of
59th Ave just west of Ontario St, on the south side of 64th Ave
west of Borden St, on the NE corner of Laurel St and 13th Ave,
and in front of a church on Kitsilano Div just east of
Macdonald St at Stephens St.

Thuja plicata D. Don
Western Red Cedar

This is one of our most common, large, evergreen trees on the
Pacific Coast and a very important timber tree, growing to 40
m tall or sometimes taller. The rusty brown stringy bark in
vertical, fibrous strips is distinctive on old trees. Tiny scale-like
leaves are yellow-green and are borne on broadly triangular,
fern-like sprays of branchlets drooping from the side branches.
Western Red Cedar looks droopier and more yellow-green than
the other local evergreens. The foliage has a strong, very
pleasant, somewhat citrus-like smell, especially when bruised.
Cones, produced in great quantity, are small, 1–2 cm long, and
made up of 4–6 flattened scales. There are two or three
flattened, winged seeds above each scale.

It is an abundant tree in the wild and is sometimes cultivated
for hedging or as a specimen shrub, quickly becoming a large
tree. Specimens may be seen throughout Stanley Park, Pacific
Spirit Regional Park, and Queen Elizabeth Park, as well as
many smaller parks. There are far fewer named cultivars of
Western Red Cedar than of its eastern North American counterpart, *Thuja occidentalis*.

'Zebrina' ('Aureovariegata') – This cultivar has bright gold
or gold and green banded foliage and is one of the best of the
golden conifers, adding brightness to our dull winter days. It
might be confused with golden cultivars of other conifers, but

the soft, slick feel of the branchlets and the very sweet fragrance are distinguishing characteristics. As with most, if not all, of the golden conifers, the yellow colour is brightest on the sunny south side of the trees. 'Zebrina' is moderately common in cultivation locally. There are two large trees on the SE corner of 15th Ave and Trafalgar St, a large one on the NE corner of 34th Ave and MacKenzie St (with a large golden *Chamaecyparis pisifera,* for comparison), and smaller individuals on the north side of Chancellor Blvd just west of Newton Cres and on the SW corner of Hemlock St and 14th Ave.

Unknown cultivar – There is a very distinctive form of *Thuja plicata* found in a number of parks in Vancouver. It has a broad columnar form, with fairly open growth and coarse, thick foliage with yellowish tips. Its name, if it has one, has not been found. There is a row on the south side of Tatlow Park, along 3rd Ave just west of Macdonald St, a row on the north side of 4th Ave between Collingwood St and Blenheim St at McBride Park, a group on the east side of Queen Elizabeth Park where the road from the Bloedel Conservatory intersects the main road through the Park, and one tree on the Main Mall in front of the Physics Bldg at UBC.

Thujopsis dolobrata (L.f.) Siebold & Zucc.
Hiba Arborvitae

A dense, broadly pyramidal shrub or small tree that grows to 15 m tall in its native Japan. The branchlets are flattened and have a very distinct plastic feeling. There is a broad white patch beneath each scale-like leaf. Variegated-leaved forms, with some fully white branchlets, are more common in cultivation than the typical all-green forms. The angular cones are a bit larger than those of *Chamaecyparis* and *Thuja* species.

It is not very commonly planted outside parks and gardens, and most of these are small specimens. There are several nice variegated ones on the UBC campus (on the south side of University Blvd at NW Marine Dr, on the NE corner of the Physics Bldg on East Mall south of the Main Library, and a large one NW of the Carriage House in Cecil Green Park), one in the lawn area on the eastern side of Queen Elizabeth Park on the road to the Bloedel Conservatory, and a group along Oak St between 33rd Ave and 37th Ave at VanDusen Botanical Garden.

Ginkgoaceae – Ginkgo Family

Ginkgo biloba L.
Ginkgo or Maidenhair Tree

A unique relative of the conifers, this native of China is a very disease-free and pollution-tolerant tree often cultivated as a street tree in cities. It is part of a much more widespread group of ancient trees, the remainder of which are all long ago extinct. Ginkgo is not known to exist in the wild but has been cultivated around Chinese monasteries for centuries. Like the true cedars (*Cedrus*) and larches (*Larix*), Ginkgo has both short thick shoots and long slender ones, giving the trees a distinctive look in winter when the branches are bare. The fan-shaped leaves have a dichotomously branched venation pattern and are like those of no other tree. The leaves often turn bright golden yellow before dropping in the autumn. Male and female trees are separate, and the females bear yellow cherry-like fruit, the flesh of which is foul-smelling, somewhat like rancid butter, so females are usually avoided as street trees. However, inside the fruit is a seed which contains an edible sweet kernel. Fresh Ginkgo seeds are seasonally available in Vancouver's Chinatown and are sometimes seen in Chinese and Japanese restaurants.

There are only a few dozen trees around the city and no really large ones. The largest ones include an individual in a street planting on the south side of 20th Ave between Oak St and Laurel St, and two on the west side of Wallace St between 39th Ave and 41st Ave. There are smaller individuals along East Blvd between 50th Ave and 51st Ave, on the west side of the Heather Pavilion at Vancouver General Hospital, on the north side of 15th Ave between Blenheim St and Waterloo St, a group at the Hydro station opposite the Catholic Cathedral at Georgia St and Richards St, several in Queen Elizabeth Park on the west side of the conservatory, and several between the pine woods and Livingstone Lake at VanDusen Botanical Garden. •5

Pinaceae – Pine Family

Abies – True Firs

The *Abies* are often called the true firs to distinguish them from a number of other trees known as firs, including, in our area, Douglas Fir (*Pseudotsuga*) and Chinese Fir (*Cunninghamia*). The needles, twigs, and cones of the *Abies* species have a unique and distinctive set of characteristics. The needles are borne individually, and are either projected laterally in two flattened rows or upward and forward on the branches. They are usually blunt-tipped and soft to the touch, compared to the more sharply pointed needles of the similar spruces (*Picea*). The twigs are smooth, with obvious large, round leaf scars after the needles drop, compared to the rough twigs of spruces. However, the major difference is that the cones are held upright on the branches and are usually only in the very tops of the trees. The cones fall apart at maturity, so that individual cone scales are found on the ground around the trees, leaving the bare cone stalk directed upward on the upper branches. Cone scales have a distinct, pointed bract attached to each. Whole cones are rarely found on the ground, only when wind has blown out a branch bearing immature cones or squirrels have removed one.

One group of firs has prominent white bands on the undersides of the needles and are known as silver firs. Several of the North American firs are collectively known to foresters as Balsam Fir or just Balsam. The true Balsam Fir (*Abies balsamea*) of north-eastern and north-central North America is not cultivated in Vancouver.

Twigs and cone
scales of *Abies*

1 *A. amabilis*

2 *A. concolor*

3 *A. grandis*

4 *A. nordmanniana*

5 *A. pinsapo*

6 *A. veitchii*

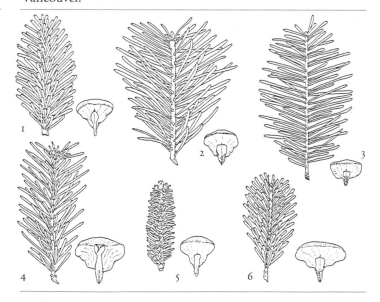

Abies amabilis (Dougl. ex Loud.) Forbes
Pacific Silver Fir or Amabilis Fir

This is the common native fir in subalpine areas of the local North Shore Mountains. It does not extend down to sea level in this part of British Columbia and is rarely cultivated here. Occurring naturally from extreme southern Alaska to northern California, it is a slow grower but may reach 70 m eventually. It needs a cool, rich soil to grow well. Examination of the branchlets under magnification shows pubescent twigs and resinous buds. The best distinguishing characteristics are the blunt or notched-tipped needles borne in two flat rows of longer needles projecting comb-like, and a third row of shorter needles pointing forward on the tops of the branchlets. The species name means beautiful and the common name refers to the silvery bands on the undersides of the needles.

The only specimens seen in Vancouver are two relatively large ones on the north side of 38th Ave between Dunbar St and Collingwood St, and small ones in the BC Native Garden of UBC Botanical Garden and the Western North American Section at VanDusen Botanical Garden.

Abies concolor (Gord.) Lindl. ex Hildebr.
White Fir or Colorado Fir

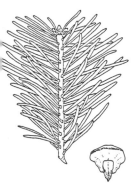

This majestic tree of the central and southern Rocky Mountains is one of the most beautiful of North American true firs. It is found in the wild from northeastern Oregon east to Wyoming and south through the Rocky Mountains to northern Mexico. The soft, grey-green needles are among the longest of any of our cultivated firs. They radiate upward and forward on the branches like a hairbrush. The colour and length of needles distinguish it from any of the other locally cultivated firs. The only other grey-green needled conifers cultivated in our area are Blue Spruce (*Picea pungens*) and Blue Atlas Cedar (*Cedrus atlantica* var. *glauca*), which might be confused with White Fir at a distance, but both of these have very sharply pointed needles. The lovely, symmetrical young trees are not seen in our gardens as often as they should be.

There is a relatively large specimen on Newton Wynd just east of Newton Cres in Point Grey, two trees on the west side of Maple St between 14th Ave and 15th Ave, two on the south side of 31st Ave just west of Elm St, two trees (one of which usually bears cones) on the west side of Blanca St between 3rd Ave and 4th Ave, and one with a large Mountain Hemlock on the east side of Mackenzie St between 19th Ave and 20th Ave. The largest specimen seen in Vancouver is on the sw corner of 3rd Ave and Waterloo St, growing with a large Monkey-Puzzle Tree. There is a very fine form with especially long (7 cm), very blue-grey needles on the NE corner of the Columbian House residence of the Vancouver School of Theology, just off Chancellor Blvd at UBC. •6

Abies grandis (D. Don ex Lamb.) Lindl.
Grand Fir

This very large tree grows to 70 m or more at low elevations along the coast from southwestern British Columbia to California. The grey bark is relatively smooth, even on older trees, when compared to that of our other local conifers. Dark green, hard needles extend out on opposite sides of the branchlets, giving them a flattened 'double-comb' appearance. The needles are flat and blunt-tipped and give off a very strong, pleasant smell of citrus when crushed. The elongated cones sit upright only on the highest branches. They are rarely seen whole on the ground, but only in pieces under the trees as they have fallen apart. It is the only true fir found at low elevations in the lower mainland, although it is not as common as most of our other cone-bearing trees. This beautiful fir is rarely cultivated locally, probably because it is native and therefore not considered desirable.

It is common in the wild in Stanley Park, Pacific Spirit Regional Park, and the University Endowment Lands. There is a large specimen along the east side of West Mall at Memorial Road on the UBC campus and some very good old specimens in the UBC Botanical Garden's Asian Garden.

Abies nordmanniana (Steven) Spach
Caucasian Fir

This very large silver fir from the Caucasus, Asia Minor, and Greece retains its branches to the ground even on older trees. The needles curve upward and forward on the branches and are a bit longer than those of the Veitch's Silver Fir, which is similar and also planted in our area. Both have characteristic white lines on the lower surface of the needles. The 10–15 cm long cones of the Caucasian Fir are twice to three times as long as those of Veitch's Fir, and the cone bracts are almost as long as the scales and often recurved. Winter buds at the tips of twigs are non-resinous, with the individual bud scales clearly visible.

The large, erect cones are borne only in the tops of mature trees, but are quite noticeable, even as bare stalks, after the cone scales have fallen.

It is not very common here. There are some lovely specimens around the city, including a large, slender one at the NW corner of Arbutus Park on SW Marine Dr just NW of Arbutus St, a large one on the SE corner of 41st Ave at Holland St, smaller specimens in the fir collection at VanDusen Botanical Garden, and one NW of the Quarry Garden and two above the tennis courts and below the Rose Garden (with two trees of *Abies veitchii* and three trees of *Abies concolor*) in Queen Elizabeth Park.

Abies pinsapo Boiss.
Spanish Fir or Bottlebrush Fir

This slender fir from Spain has distinctive, very stiff (but not sharp) needles, borne all around the twigs in a bottlebrush fashion. It is very pleasing to the touch. The needles of the typical form are dark grey, but the form more commonly seen in cultivation is the paler grey cultivar 'Glauca.' The tree is said to be becoming rare in nature, but is often seen in collections. It is rarely grown here but is so distinctive that it will be remembered once seen or, especially, felt.

There are two grey-needled specimens in the lawn area between the medical buildings north of Woodward Library at UBC, a very good specimen at the south entrance to the dining hall of Place Vanier Residences at UBC, a green form and two grey ones west of the Heather Garden in VanDusen Botanical Garden, and a grey form on the SE corner of 19th Ave and Crown St.

Abies veitchii Lindl.
Veitch's Silver Fir

This pyramid-shaped Japanese fir has needles that are bright green above, but which appear very silvery when viewed from below because each needle has a pair of white lines on the underside. They are very soft and plastic-feeling to the touch. The needles are somewhat parted along the upper side of the twigs, forming a V-shaped trough. Winter buds are rounded and purple with a thick coat of resin that hides the individual bud scales. The twigs are usually much hairier than those of the similar *Abies nordmanniana*, and the cone bracts are much shorter than the scales.

It is not very common here. There are several symmetrical specimens on the SE corner of 49th Ave and McKinnon St, on the NE corner of 33rd Ave and Culloden St, on the north side of 54th Ave just east of Rupert St; there is a large one planted behind two smaller specimens of *Abies concolor* on the west side of Maple St between 14th Ave and 15th Ave, one on the NW side of Totem Park Residences at UBC, and two large specimens (with two of the similar *Abies nordmanniana* and three greyer *Abies concolor*) above the tennis courts (below the Rose Garden) in Queen Elizabeth Park.

Cedrus – True Cedars

These Old World conifers grow to be very large and picturesque trees with stiff evergreen needles on two types of twigs. The needles on short twigs appear as dense tufts, while those on long shoots toward the ends of the branches are sparse and scattered, spiralled along the twigs, and generally longer than those of the short shoots. Locally, the only other cone-bearing trees with a similar needle arrangement are the larches (*Larix*) which have very soft, deciduous needles. The cedars have very fat, rounded female cones borne upright on the branches and falling apart at maturity to leave only the spike-like stalk on the branch. Male cones are long, slender, and often produced in great abundance, but are of short duration. The true cedars are difficult to distinguish from one another, although some characteristics have been included under each species. Some authorities have gone so far as to say that they all should be considered one variable species. One authority on trees says that it is a brave person who is able to always distinguish between the different species. And, Alan Mitchell, one of Britain's leading tree authorities, says that he cannot always distinguish between them, especially young trees of Cedar-of-Lebanon and the green forms of Atlas Cedar.

Cedrus atlantica (Endl.) Manetti ex Carr.
Atlas Cedar

A large pyramidal tree growing to about 40 m tall in the Atlas Mountains of North Africa, Atlas Cedar has long been a popular park and garden tree. The dark green needles are stiff and shorter than those of *Cedrus libani* or *Cedrus deodara*. The green form is very difficult to distinguish from typical *Cedrus libani*, but it is more commonly cultivated here than the latter,

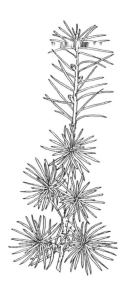

and the ends of the branches and top are held upright. Needles are slightly more uniform in length and they are slightly shorter on average (1–1.5 cm) than those of *Cedrus libani.*

The green form is less frequently planted than is the next variety. There are a number of specimens around town, including a large one on the SE corner of Angus Dr at 49th Ave, two large ones on Kingsway at Moss St (in front of the Purdy's Bldg), and one in Stanley Park north of the perennial beds near the Rose Garden.

Cedrus atlantica var. *glauca* Carr., Blue Atlas Cedar – This wild form is considered more desirable as a garden tree because of the clear blue-grey colour of its needles. It is far more commonly cultivated around the city. There are many magnificent specimens, including those on the SE corner of Granville St and 49th Ave, on 41st Ave between Marguerite St and Churchill St, at Devonshire Cres and Selkirk St, on the south side of Cedar Cres between Burrard St and Fir St, and one espaliered against the front wall of the Vancouver Parks Board Office on Beach Ave.

'Pendula,' Weeping Blue Atlas Cedar – This blue-grey cultivar has a very drooping form that will spread along the ground unless grafted on an upright trunk or staked upright until it becomes woody. There is a beautiful specimen just outside Sprinklers Restaurant and a smaller one in the Mediterranean Section at VanDusen Botanical Garden, one between the Alpine Garden and the Winter Garden in UBC Botanical Garden, and one planted so that it hangs down a wall on the SW corner of 8th Ave and Willow St.

Cedrus brevifolia (Hook.f.) Dode (*Cedrus libani* A. Rich. var. *brevifolia* Hook.f.) Short-Needled Cedar or Cyprian Cedar

A native of Cyprus, this tree is either treated as a distinct species or as a variety of the Cedar-of-Lebanon. The distinguishing feature is the very short needles (about 1 cm long on the short shoots). The needles are very dark green.

The rarest of the true cedars to be found locally, but it is very distinctive. There is a large one in the Old Arboretum on the UBC campus, some young ones in the Mediterranean Section in VanDusen Botanical Garden, and one on the north side of the Pitch & Putt Golf Course in Stanley Park.

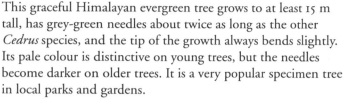

Cedrus deodara (D. Don) G. Don
Deodar Cedar or Himalayan Cedar

This graceful Himalayan evergreen tree grows to at least 15 m tall, has grey-green needles about twice as long as the other *Cedrus* species, and the tip of the growth always bends slightly. Its pale colour is distinctive on young trees, but the needles become darker on older trees. It is a very popular specimen tree in local parks and gardens.

There are many old trees around the city, including one on the west side of Knight St between 37th Ave and 39th Ave; two in front of the Vancouver School of Theology on Iona Dr at UBC; and large ones at the Aberthau Cultural Centre at Trimble St and 2nd Ave, on the SE corner of 19th Ave and Puget Dr, and the SE corner of 21st Ave and Puget Dr. There are particularly weeping forms with sparse, open growth along the west side of Granville St between 46th Ave and 47th Ave, on the east side of Granville St between 34th Ave and 35th Ave, and on the east side of Yew St north of 53rd Ave.

'**Aurea,' Golden Deodar Cedar** – This form has golden needles, especially on younger growth and on the south side of trees where the foliage receives the most sun. There are two trees on the east side of Chaldecott St between 27th Ave and 29th Ave, two on the grounds of David Thompson Secondary School, and several by the waterfall in the Sino-Himalayan Garden at VanDusen Botanical Garden.

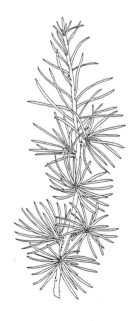

Cedrus libani A. Rich.
Cedar-of-Lebanon

This famous evergreen of the Middle East becomes a large, spreading, picturesque tree. It is a familiar sight as an old specimen tree on the grounds around European cathedrals and on large estates. The needles are dark dull green, the colour of the green forms of *Cedrus atlantica,* but those of the Cedar-of-Lebanon are a bit longer (1.5–2 cm) and they are more variable in length within one branch cluster. The twigs of Cedar-of-Lebanon tend to be glabrous or very slightly short-hairy, compared to the very hairy twigs of the Atlas Cedar, but this is a variable characteristic. One almost has to have branches from both trees in hand to be able to distinguish the two.

This is surprisingly rare in parks and gardens here, although there are a few small ones around the city. There is a very nice specimen that was planted in 1954 in Queen Elizabeth Park south of the Quarry Garden, and two on the sw side of the Pitch & Putt Golf Course in Stanley Park.

Larix – Larches

The *Larix* species are usually known as larches, or some as tamaracks, and are the only North American members of the pine family that are deciduous. Like the true cedars (*Cedrus*), the larches have two types of shoots: short, thick ones along the older parts of the twigs that bear dense, spirally-arranged tufts of needles, and long slender shoots at the tips of branches that bear loosely scattered, usually longer needles. The short shoots continue to grow each year, but at a very slow rate, so they remain thick and knob-like. Unlike the evergreen needles of the true cedars, those of the larches are soft and pale green, turning golden yellow in the autumn before they drop. Smaller branches of most of the larches are very drooping. Soft female cones emerge from the short shoots in early spring. They are usually bright red or purple and are noticeable from a distance. Rounded tufts of ephemeral, yellow male cones are produced at the same time on other short shoots. After pollination, the female cones mature and become pale brown, usually remaining on the trees for many years.

Larix decidua Mill. (*Larix europaea* DC.)
European Larch

This is the common European species of larch, which is sometimes cultivated here. It may be distinguished from the other similar large-coned larch species by cone scales that are straight or slightly incurved at the tips. The young twigs are a distinctive yellow when compared to other species. It is an important source of timber and turpentine in Europe.

There is a large grove on the NW side of Queen Elizabeth Park (incorrectly labelled as *Larix occidentalis*), some just north of the Parks Board Office in Stanley Park (with *Larix occidentalis* and *Larix kaempferi*), a large one along Main Mall in front of the Chemistry Bldg and one in the Old Arboretum at UBC, and two trees west of the Perennial Garden in VanDusen Botanical Garden.

Larix × *eurolepis* A. Henry
(*Larix kaempferi* [Lamb.] Carr. × *Larix decidua* Mill.)
Hybrid Larch

This vigorous hybrid between the common European and Japanese species arose in an English garden around the turn of this century and it combines the characteristics of both species. It is not always easy to distinguish from the Japanese Larch (*Larix kaempferi*), although the hybrid has cones that are slightly more elongated and the tips of the cone scales are less reflexed than the parent. The hybrid name is of some interest as it combines *europaea* and *leptolepis,* the names by which the parent trees were known at the time the hybrid was named. Both the parent species are now known by other names, but the hybrid name must remain unchanged, according to the rules of botanical nomenclature.

There are a number of trees between the parking lot and the entrance at VanDusen Botanical Garden, and one in the Winter Garden at UBC Botanical Garden. •7

Larix kaempferi (Lamb.) Carr.
Japanese Larch

This Japanese species has needles that are slightly more bluish green with more prominent bands of white stomata on the under-sides than the other species cultivated locally. Its cones are more rounded in outline and they have more reflexed scales than those of other similar species. The smaller twigs have a definite reddish cast.

There is one north of the Parks Board Office in Stanley Park (with *Larix decidua* and *Larix occidentalis*), a street planting along the east side of Elliott St between 54th Ave and 56th Ave (with *Larix occidentalis*), a tree at the western end of the Rhododendron Walk at VanDusen Botanical Garden, and a small grove of trees on the west side of MacMillan Bldg at UBC.

Larix laricina (Du Roi) C. Koch
Tamarack

A common, deciduous, cone-bearing tree across the northern parts of Canada and into the northern United States, the Tamarack is usually found in the edges of boggy sites. In British Columbia, it is found in the central and northern parts of the province. It cannot be confused with any of the other larches when the cones are present as the cones are less than 2 cm long, or less than half the size of the others. The twigs are used in crafts, especially in making the familiar twig Canada geese.

It is rarely cultivated locally. There is a relatively large tree in the Old Arboretum at UBC, and a grove of young trees in the Eastern North American Section and a tree in the Canadian Heritage Garden at VanDusen Botanical Garden.

Larix occidentalis Nutt.
Western Larch

This is one of three native *Larix* species in our province. Western Larch is found in those mountains east of the Cascade Mountains and south to Oregon. The pale green needles in spring and summer and the golden yellow autumn colour make it easily recognizable in nature. It is not cultivated as often as other *Larix* species, from which it may be distinguished by the cones bearing long, slender, usually recurved, bracts extending beyond each cone scale. None of the other locally cultivated larches have bracts as long as the scales. The third native species, *Larix lyallii,* is found at higher elevations in British Columbia and is not in cultivation, except in the BC Native Garden at UBC Botanical Garden.

There is a small tree in Stanley Park just north of the Parks Board Office (with *Larix decidua* and *Larix kaempferi*), a street planting along the east side of Elliott St between 54th Ave and 56th Ave (mixed with *Larix kaempferi*), and some young ones in the BC Native Garden at UBC Botanical Garden.

Picea – Spruces

Spruces are among the dominant trees of the northern reaches of tree life. Often confused with firs (*Abies*), the short needles of spruces are often square in cross-section, are usually very sharply pointed, and are borne singly but in dense spirals around the twigs, like a bottlebrush. Each needle is produced on a short peg which is left behind on the twig after the needles have fallen, thus leaving the twigs with a distinctive rough texture. Pendulous cones lack a visible bract, characteristic of the firs, and they remain intact at maturity, not falling apart as do those of the firs. There are some forty-five species in the genus.

Picea cones

1 *P. abies*
2 *P. breweriana*
3 *P. glauca*
4 *P. omorika* and seed
5 *P. orientalis*
6 *P. sitchensis*
7 *P. pungens*
8 *P. torana*

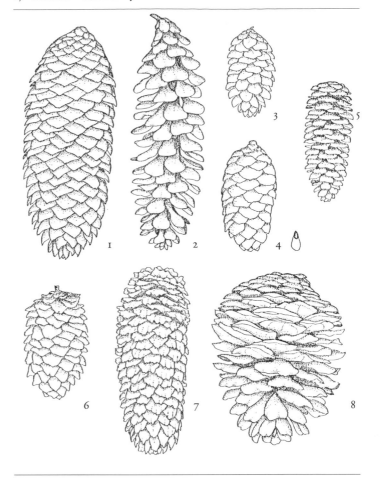

Picea abies (L.) Karst.
Norway Spruce

This evergreen tree of central and northern Europe is the most commonly cultivated spruce in gardens. It is seen as the typical wild form as well as in many horticultural, often dwarf, forms. It is often used as a small plant in foundation plantings in front of homes and buildings, but it quickly outgrows its location. On old trees, the long drooping secondary branches hanging from the upturned whorl of main branches is very characteristic. The pendulous cones are the largest of the spruces, growing to 10–15 cm long, with irregularly notched scales. Dark green needles (1–2 cm long) are very sharp-pointed and prickly to the touch. The growth habit and needles are very much like that of Oriental Spruce, but both the needles and cones of Norway Spruce are longer.

There are many specimens planted in the city. The largest trees, with the characteristic branching pattern, are found on the sw corner of Angus Dr and 33rd Ave, on the SE corner of Connaught Dr and Avondale Ave, in McCleery Park at Marine Cres and 49th Ave (two trees), and east of the Rose Garden in Stanley Park.

Picea breweriana S. Wats.
Brewer's Weeping Spruce

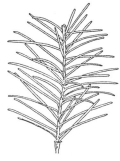

One of the rarest spruces in nature, Brewer's Weeping Spruce is found only on a few mountain passes in the Siskiyou Mountains of southwestern Oregon and northwestern California. The sight of mature specimens of this magnificent tree in nature is an unforgettable experience for anyone loving trees. It is a much sought-after spruce for gardens and arboreta in temperate climates around the world. The ultimate branches weep for great distances from the sweeping lateral branches. The grey-green needles are long and soft, curving outward and upward from the branches. The cones, although rarely produced locally, are long (7–10 cm). It is a very distinctive tree, unlike any other spruce in cultivation in Vancouver. It is generally unavailable from nurseries and seems to be fairly slow to get established in gardens, making it all the more desirable.

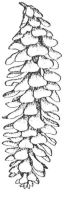

There are small specimens in the Alpine Garden at UBC Botanical Garden, in the edge of the Pine Woods at VanDusen Botanical Garden, and on the slopes NW of the Bloedel Conservatory in Queen Elizabeth Park. The finest specimen in the city, large enough to bear cones, is on the east side of Angus Dr between 36th Ave and 37th Ave.

Picea glauca (Moench) Voss
White Spruce

A blue-green spruce with sharply pointed needles less than 2 cm long. The needles are more delicate, about half as long, and not as sharp as those of the similar Blue Spruce (*Picea pungens*), nor is it as blue-green. White Spruce is native across the interior of Canada and the northern part of the United States. It is very cold-tolerant and is frequently cultivated as a garden plant in colder areas. It is not seen very often along the coast, where it does not grow very well.

There is a row of trees along the NE side of Westmount Park along Drummond Dr at Blanca St (although they are not very lush), several small trees near the large grove of larches and with the similar *Picea abies* on the NW side of Queen Elizabeth Park, and small ones in the BC Native Garden at UBC Botanical Garden.

Picea omorika (Panc.) Purk.
Serbian Spruce

Superficially, this tree looks very much like a narrow, drooping Douglas Fir. It is a narrow, pyramidal tree with short weeping side branches that curl up at the tips, which grows to about 30 m tall in nature in southeastern Europe. The needles appear to be dull green with whitish undersides but are twisted on the branches so that the upper surface is really grey and the lower dark green if the twig is held upright. The mature cones are bright brown, 4–6 cm long, and are usually borne in the tops of the trees. Trees here often do not produce many cones.

It is not a very common tree here, but very distinctive and recognizable. There is a specimen on the north side of 33rd Ave just east of Oak St, a large one on the sw corner of 13th Ave and Cypress St, several nice specimens in the Conifer Collection at VanDusen Botanical Garden, five in a yard on the south side of 54th Ave just east of Cartier St, four on the west corner of Broughton St and Barclay St, and a number along the entrance to Shaughnessy Golf and Country Club on sw Marine Dr near 41st Ave. One of the tallest specimens is against a house on the NW corner of Blenheim St and 22nd Ave.

Picea orientalis (L.) Link
Oriental Spruce

This large tree is native to western Asia and closely resembles the habit of Norway Spruce, but is much less common in cultivation. It grows rapidly and is known to reach 50 m tall. The bright green needles are shorter (to 1 cm long) with blunt tips so that the branchlets feel much softer than those of Norway Spruce. The overall habit of the Oriental Spruce is less weeping and of a more compact growth, and the cones are smaller (6–10 cm) with more rounded scales.

It is not often grown here. There is a large tree in the Old Arboretum at UBC, one in The Crescent at Hudson St (with several Norway Spruce for comparison), a beautiful old specimen with many cones in the median of Angus Dr between Hosmer Ave and Granville St, a pair on the NE corner of Hudson St and Devonshire Cres, and a very large one that bears cones prolifically on the NE corner of Highbury St and 23rd Ave.

Picea pungens Engelm.
Blue Spruce or Colorado Spruce

This familiar tree is a pyramidal evergreen, native to the central Rocky Mountains, that has very sharp needles and bright yellowish brown cones with scales notched at the tips. The cones are usually borne only in the very tops of old trees and are produced only sporadically in Vancouver. In nature, it is extremely variable in the colour of the needles and in the habit of the tree. Most of the specimens cultivated have been selected for the blue or grey colours of the needles, hence the common name of Blue Spruce, but there are also many dwarf and creeping forms that have been selected.

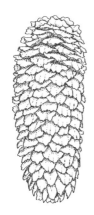

It is commonly cultivated as a specimen plant in yards and sometimes as a street tree in the city. There are good blue specimens on the grounds of Point Grey Secondary School at East Blvd and 37th Ave, a very blue, tall specimen on the NW corner of Pine St and Cedar Cres, one on the NE corner of 47th Ave and Cartier St, two on the SW corner of 6th Ave and Sasamat St, and a number of nice specimens along the drive to the Shaughnessy Golf and Country Club off SW Marine Dr near 41st Ave. The greener form is not considered as attractive and is grown less often, but there are specimens (alternating with Bigtrees, *Sequoiadendron*) along the median of Cambie St between King Edward Ave and 29th Ave, an old specimen in the Old Arboretum at UBC, and many specimens (obviously seedlings and variable in habit and colour, but mostly dark green) in the median of King Edward Ave from Dunbar St to Quesnel Dr. •2, 8

Picea sitchensis (Bong.) Carr.
Sitka Spruce

A massive evergreen timber tree that grows along the coast from Alaska to northern California but never very far inland. Its most distinctive characteristic is the extremely sharp-pointed, slender needles, especially on young specimens. The needle colour is usually dull green or grey-green. The needles are more slender and duller than the similar Blue Spruce.

It usually is not very attractive when cultivated and is not often grown here. There are a number of trees, probably natural, along the fairways of Point Grey Golf Club, along 51st Ave at Dunbar St, and other areas of Southlands in south Vancouver, and frequently in Pacific Spirit Regional Park along SW Marine Dr between 16th Ave and 41st Ave. There is a large old specimen in the Old Arboretum at UBC, two relatively large trees,

with limbs to the ground, along the north side of the main drive through Queen Elizabeth Park, and a large one on the north side of 52nd Ave between Laburnum St and Cypress St. •9

Picea torana Koehne
(*Picea polita* [Siebold & Zucc.] Carr.)
Tigertail Spruce

This very distinctive Japanese spruce grows to about 30 m tall in nature. The bright green, 2 cm long needles are very stiff and sharp-pointed (stiffer and sharper than those of the similar Norway Spruce) and radiate out all around the bright yellow-brown twigs. Winter buds are large and reddish brown, and the previous year's bud scales remain as a prominent dark ring on the branch at the base of the new growth. The cones are large (10–15 cm long), somewhat like a short, thick Norway Spruce cone.

The only tree seen in Vancouver is a large one with cones in a yard on the NW corner of King Edward Ave at Hudson St.

Pinus – Pines

These familiar evergreen trees have needles borne in distinct groups (fascicles) of twos, threes, or fives in our species. Those with needles in clusters of fives are known as the white pines. Their needles are slender, very flexible, soft to the touch, and have fine white lines along their length, giving them their common name. The black or red pines (or hard pines) have stiffer, thicker needles in bundles of twos or threes. Pines are among the most common of our temperate conifers, both in nature and as garden trees.

Pinus aristata Engelm.
Bristle-Cone Pine

There are only a few small individuals of this tree in Vancouver, but it is included because it is of great botanical interest. Bristle-Cone Pines occur at very high elevations from California to Colorado and are considered to be among the oldest living organisms. They are relatively fast-growing when young but eventually grow extremely slowly and never reach more than about 15 m tall. Some individuals are known to be around 5,000 years old. The dark blue-green needles are only 2–4 cm long and are held in tight clusters of five, usually lying close to the limbs. The needles bear distinctive white resinous dots that are often mistaken for scale insects. The relatively small (6 cm

long), egg-shaped cones are pale brown and lightweight, with
very slender, sharp tips to their scales. The cones are also
covered with white resin.

There is a tree about 3–4 m tall in a garden on the south side of
41st Ave at sw Marine Dr, one in the Dwarf Conifer Collection
at ubc Botanical Garden, and a number at VanDusen Botanical
Garden (including two in the Pine Collection near the end of
Livingstone Lake, one in the Alpine Garden, and several, with
cones, in the Dry Garden).

Pinus armandii Franch.
Armand Pine

This graceful 5-needled pine is native to China, Burma, and
Formosa. The long needles (10–20 cm) usually droop from the
limbs and some of them have a strong bend near the base or
about half-way out. The cones are longer than they are broad
(10–16 cm long and 6–8 cm broad) and are borne on stalks 3 cm
long. The cone scales are thicker than those of our other 5-
needled pines. This tree is very much like the Himalayan White
Pine (*Pinus walliachiana*), except the cones are shorter and
broader and the seeds larger and wingless. The seed of the
Himalayan White Pine has a flattened wing about twice as long
as the seed itself. The two are difficult to distinguish until cones
are produced, although Armand Pine has slightly more
yellowish green needles when the species are seen together.

It is rare locally. The only ones found are in a recent planting
with Eastern White Pine (*Pinus strobus*) and other pines on the
south side of Granville Island, and five young trees at the end of
Bentall St (north of Grandview Hwy, one block east of Rupert
St). These have produced cones in recent years, so their identity
could be confirmed.

Pinus banksiana Lamb.
Jack Pine

A common pine of the northeastern United States and much of
Canada, extending as far west as northeastern British
Columbia. The needles, in clusters of twos, are the shortest of
our pines (only 2–3 cm long) and distinctly yellow-green, much
like those of the Shore Pine, except shorter. The cones are
gnarled and twisted, 2–5 cm long, and curved closely against
the branches. They normally remain closed for many years. It is
not a particularly attractive tree and is more often planted as a
curiosity or as part of a conifer collection than for its beauty.

It is rare here. There are three nice specimens (planted with

Pinus ponderosa and *Pinus sylvestris*) on the slopes along a paved path north of the Bloedel Conservatory in Queen Elizabeth Park, and a large one in the Old Arboretum at UBC.

Pinus contorta Dougl. ex Loud.
Shore Pine and Lodgepole Pine

This is our only locally native pine, found on rocky coastal slopes above the ocean or in elevated areas of coastal bogs. The trees are often gnarled or twisted on rocky outcrops, but they may be straighter when growing under more ideal conditons. The 3–5 cm long needles are in twos, twisted, and yellowish green. The cones are 3–5 cm long and may or may not remain closed for a number of years before releasing their winged seeds. There is a sharp prickle on the end of each cone scale.

The native Shore Pine is *Pinus contorta* var. *latifolia,* which differs largely in its growth habit from the very slender, straight Lodgepole Pine (*Pinus contorta* var. *contorta*) found so abundantly in the interior of western North America.

It is not often cultivated here. There are a number of native trees around the city, including a grove of trees at the Langara Golf Course on the east side of Cambie St from 49th Ave to 58th Ave, a grove on the west side of Crown St at the western end of King Edward Ave, and a grove referred to as the Pine Woods in VanDusen Botanical Garden.

Pinus coulteri G. Don
Coulter Pine or Big-Cone Pine

This long-needled pine grows only on mountain slopes along the coast of California from about San Francisco south to Baja California. Long, dark bluish green needles (15–30 cm long) are borne in threes and arch from the twigs. The cones are the largest and heaviest of all pines, often growing to 40 cm long, and are bright golden brown in colour with very long stout prickles on the cone scales. They are borne flat against the trunk and main limbs, and are usually asymmetrical. In nature, even relatively small trees bear cones.

There are probably fewer than half a dozen trees in the city, mostly at UBC. Probably the largest individual (which is beginning to bear a few, relatively small cones) is on the north side of Selkirk Elementary School at 22nd Ave and Welwyn St. Specimens at UBC include one in the Old Arboretum, one on the south side of University Blvd at NW Marine Dr, one on the NW corner of Main Mall and University Blvd (in front of Scarfe

Bldg), and one in the courtyard of MacMillan Bldg off Main
Mall. There is also a small tree on the west side of West Blvd
just south of 54th Ave.

Pinus densiflora Siebold & Zucc.
Japanese Red Pine

This large Japanese native grows to 35 m tall in nature and is
not often cultivated in our area. It is fairly nondescript and can
be confused with some of our more commonly cultivated pines.
The needles, borne in pairs, are very slender and are brighter
green and somewhat less twisted than those of the similar *Pinus
sylvestris* and *Pinus contorta*. The cones are about 5 cm long.
There are a number of variegated, weeping, and dwarf cultivars
of Japanese Red Pine.

There is a large, typical wild form on the hilltop on the east side
of the Heather Garden at VanDusen Botanical Garden, one on
the east side of Macdonald St between 14th Ave and 15th Ave,
and two on the west side of Main Mall at the Frederic Wood
Theatre and one on the west side of Lower Mall at Place Vanier
Residences at UBC.

'Umbraculifera,' Tanyosho or Japanese Umbrella Pine – This
dense, rounded form is the cultivar often grown in gardens for
its unusual growth habit. It usually bears cones prolifically.
There is one on the east side of the Heather Garden at
VanDusen Botanical Garden, one on the north side of the
Faculty Club at UBC, and a pair in a garden on Kevin Pl off
29th Ave.

Pinus jeffreyi Grev. & Balf.
Jeffrey Pine

This attractive pine, found in southern Oregon and throughout
California, is very similar to the familiar Ponderosa Pine and is
probably derived from it, but has adapted to more severe conditions. It is usually not as large and the needles are longer, stiffer,
and slightly greyer. The two are easily separated when cones are
present, as those of Jeffrey Pine are 15–25 cm long, at least twice
the size of those of Ponderosa Pine.

There is a nice grove in the Dry Section near the Sino-
Himalayan Garden at VanDusen Botanical Garden, a grove on
the sw side of Queen Elizabeth Park, and a row on the north
side of 57th Ave west of Cambie St.

Pinus koraiensis Siebold & Zucc.
Korean Pine

This pine is a native of Japan and Korea, growing to 30 m tall or more. Needles are of variable length, even on the same twigs, but average about 6 cm. They are borne in groups of fives, are dark grey-green, slender, and slightly bent or twisted, although less so than on the similar Japanese White Pine. The cones are oblong and about 10 cm long, or about twice as long as those of the Japanese White Pine. Cones are produced infrequently on the few trees in our area.

It is very rare here. The largest ones are on the north side of 10th Ave at Wallace Cres and on the sw corner of 47th Ave and Cartier St, and there are two smaller, slender trees in the median of Angus Dr between Granville St and Hosmer Ave.

Pinus monticola Dougl. ex D. Don
Western White Pine

A common, native, 5-needled pine of the Pacific Northwest, found from British Columbia to California. It closely resembles the Eastern White Pine (*Pinus strobus*) but is a narrower tree with denser growth, slightly stiffer needles, and larger cones (to 25 cm long). The long, slender cones are relatively light for their size and hang on stalks from near the tips of branches. The cone scales are flattened and lack any prickles on the tips. The trees have a soft, grey-green look at a distance. Close inspection of the needles shows a thin white line running the length of each needle. Young trees without cones are difficult to distinguish from the Eastern White Pine.

The largest ones in the city are in Stanley Park just north of the Rose Garden and the perennial beds (behind a large golden form of Lawson Cypress), on the south side of 40th Ave between Granville St and Churchill St, on the west side of Blanca St between 5th Ave and 6th Ave, on Acadia Rd north of Chancellor Blvd at Kingston Rd, and on the south side of Toronto Rd just west of Acadia Rd. •10, 11

Pinus mugo Turra
Swiss Mountain Pine or Mugo Pine

This small tree or sprawling shrub is usually seen in our gardens as a dwarf mounded shrub, but it can be a relatively large, single- or multi-trunked tree, often of picturesque shape. The bright green needles are in clusters of twos, 3–6 cm long, and usually have a slight twist and curve. The cones are among the smallest of our cultivated pines, being 3–6 cm long and about as wide, slightly asymmetrical, with cone scales that are usually smooth and rounded at the tip. The cones are dull brown on the outside and bright golden brown inside after they open to release the seeds.

There are a few relatively large specimens in the city, including two tall single-trunked trees (very atypical of the usual forms seen here) on the north side of the Plant Sciences Greenhouse at UBC, and relatively large trees on West Blvd just north of 57th Ave, on the north side of 45th Ave between Carnarvon St and MacKenzie St, and at the top of Almond Park along Alma St at 14th Ave. Probably the best specimen in Vancouver is a very picturesque, 5-trunked old tree between 1st Ave and Point Grey Rd at Blenheim St with a smaller, twin-trunked specimen nearby.

Pinus nigra Arnold
Black Pine or Austrian Pine

This very dark green pine grows to a height of 30 m or more in its natural habitat in Europe and western Asia. The relatively stiff, thick needles (to 15 cm long) are borne in clusters of twos. Light brown cones (to 8 cm long) either lack prickles or are very slightly prickly at the base, and are borne in profusion, even on relatively young trees. The very similar Japanese Black Pine (*Pinus thunbergii*) is rarely grown here, but the two may be easily confused.

It is our most commonly planted pine, although most specimens are relatively young trees. The largest are probably ones on the NE corner of 14th Ave and Trimble St, on the SW corner of 37th Ave and Marguerite St, and south of the Rose Garden in Stanley Park. There are also large ones on Point Grey Rd at Blenheim St, two on the SE corner of Trafalgar St and 1st Ave, and one at Selkirk St just north of King Edward Ave. It is commonly planted on the UBC campus, including a number of different-sized ones between the Education Bldg and Commerce Bldg along University Blvd just west of Main Mall.

Pinus parviflora Siebold & Zucc.
Japanese White Pine

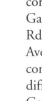

This is a very graceful and beautiful, small, 5-needled pine, growing to about 15 m tall in its native habitat in Japan. The needles are bluer, shorter (4 cm long), and more curved than the other 5-needled pines cultivated here. Cones, 6–8 cm long, are often produced on young trees.

There are few specimens in the city. There is one on the west side of the Vancouver Parks Board Office at Beach Ave and Park Lane, several in the Alpine Garden and Pinetum at VanDusen Botanical Garden, and several in the Alpine Garden of UBC Botanical Garden.

Pinus ponderosa Dougl. ex P. Laws. & C. Laws.
Ponderosa Pine or Western Yellow Pine

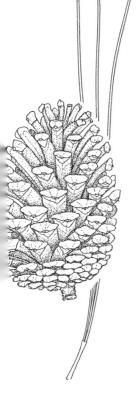

This is the common pine of the dry interior of western North America, growing from central and eastern British Columbia south to Baja California. The trees often grow separated from one another, giving the feeling of parkland. Stiff, dull green needles, 12–25 cm long, are borne in groups of threes. The large, oval cones (to about 12 cm long) are pale brown and have sharp prickles on the ends of each scale. In dry areas, the bark of old trees becomes separated into distinctive and very attractive cinnamon brown plates. There is a definite fragrance of vanilla when the cracks of the bark are smelled. However, the bark remains grey and atypical here on the wetter coast and lacks the fragrance. The name comes from ponderous, that is, having great weight or mass, referring to the size of the trees.

There are a number of relatively large trees around the city, including specimens on the north side of 17th Ave just west of Cedar Cres, one in Stanley Park along Park Lane at the west end of Nelson St just east of the tennis courts, and (at UBC) one along West Mall in front of the Ponderosa Cafeteria and one at the NE corner of the Totem Park Residences at West Mall and Agronomy Rd.

Pinus rigida Mill.
Pitch Pine

A common, eastern North American conifer, Pitch Pine grows from New Brunswick to Georgia. The slender, 6–10 cm long needles are bright yellowish green and borne in clusters of threes. Cones are about 5–8 cm long, with a sharp prickle on the tip of each cone scale, and they often remain attached to the branches for many years. A distinctive characteristic of the tree is the short, tufted twigs that arise from the trunk and main branches.

It is very rare. The only remaining tree in the city is a very large specimen in the Old Arboretum at UBC. A small one in the traffic island at Park Lane and Nelson St just east of the tennis courts in Stanley Park was removed in the winter of 1990–91, after it was damaged by snow.

Pinus strobus L.
Eastern White Pine

This is the only 5-needled pine native to eastern North America. It is obviously closely related to our Western White Pine, and the two are not always easy to distinguish in cultivation. The needles are slightly softer than our native Western White Pine and the cones are shorter (12–14 cm long).

There is a large specimen in the Old Arboretum, one on the south side of University Blvd at NW Marine Dr, and two smaller ones on the west side of West Mall at Crescent Rd, all on the UBC campus; there are several recently planted trees on the south side of Granville Island (growing with *Pinus armandii*, which has longer, more drooping needles).

'Fastigiata' – An unusual form with upright branches that form a broad columnar crown. There are two young trees on the NW corner of 16th Ave and Blenheim St, and a row on the south side of the lane between Broadway and 10th Ave east of Burrard St.

'Pendula' – An upright cultivar that has very pendulous branches. There are three trees on the lane between 12th Ave and 13th Ave east of Oak St, and a nice one in the Eastern North American Section at VanDusen Botanical Garden.

Pinus sylvestris L.
Scots Pine or Scotch Pine

This attractive, large pine from Europe and Asia has needles in clusters of twos. They are usually twisted, often bluish or greyish green, and relatively thicker for their length than those of other similar pines. Trees with the greyest needles have been particularly selected for cultivation. A unique characteristic of old specimens is the very coppery-coloured bark. The limbs of old specimens are often gnarled and twisted, making them very picturesque.

It is one of the two or three most commonly planted pines in the city and there are many small specimens. Among the larger ones are three trees in Arbutus Park at Arbutus St and SW Marine Dr, several including a very tall one with characteristic coppery bark on the upper trunk and main limbs in Memorial South Park off 41st Ave, several moderately old specimens in West Point Grey Park on the north side of 8th Ave at Trimble St, a number of specimens (with the much darker green *Pinus nigra*) in the median of King Edward Ave between Macdonald St and Vine St, and (at UBC) a group on the west side of Wesbrook Mall just north of University Blvd (by the General Services Administration Bldg).

Pinus thunbergii Parl.
Japanese Black Pine

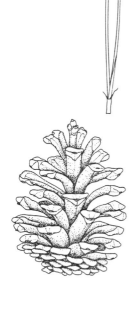

This is a commonly grown black pine in some parts of the world, especially in areas within the reach of ocean salt spray. However, there are very few trees locally, and these are mostly young, small ones. In Japan, its native habitat, the trees grow to over 30 m tall. It is a 2-needled pine that is often difficult to distinguish from Austrian Black Pine (*Pinus nigra*) unless the two are growing side by side. Japanese Black Pine needles are slightly shorter (8–12 cm long), stiffer, more sharply pointed, and not quite as dark green. The over-wintering slender vegetative shoots, known as 'candles,' are generally whiter and the cones are also smaller (about 6 cm long) on the Japanese Black Pine. There are a number of dwarf forms cultivated.

Unlike the ubiquitous Austrian Black Pine, the only specimens of Japanese Black Pine in the city seem to be in public gardens. Among the larger ones are several in front of the Asian Centre on the UBC campus, in Nitobe Memorial Garden of UBC Botanical Garden, and in the Dr. Sun Yat-Sen Classical Chinese Garden and Park in Chinatown.

Pinus wallichiana A.B. Jacks (*Pinus griffithii* Parl)
Himalayan White Pine

One of our most beautiful and graceful 5-needled pines is native to the Himalayas. The long needles (12–20 cm) are blue-green in colour, droop from the branches, and are often bent at the base or toward the middle, especially those near the ends of vigorous young branches. The slender, pendulous cones are 15–30 cm long and are borne on stalks 2–4 cm long. The pale brown seeds, about 0.5 cm long, have a wing 1–2 cm long. This characteristic separates it from the very similar *Pinus armandii,* which bears wingless seeds. The tree is border-line hardy here, and some young trees were killed or damaged during some recent cold winters.

There is a relatively large one (which bears cones in some years) on the east side of Crown St between 28th Ave and 29th Ave (nearly hidden behind a Blue Atlas Cedar and a Sawara Cypress), one below Queen Elizabeth Park on the west side of Yukon St between Nigel Ave and Talisman Ave, one on the SE corner of 54th Ave and Adera St, three at the Lumberman's Arch in Stanley Park, and a number in the Pinetum and Sino-Himalayan Garden at VanDusen Botanical Garden. There was a beautiful specimen on the UBC campus on the south side of

University Blvd just east of NW Marine Dr, planted among a number of other conifers, that was removed recently.

Pseudotsuga menziesii (Mirb.) Franco
Douglas Fir

This is one of *the* common, important, evergreen timber trees of the Pacific Northwest, growing naturally from southwestern British Columbia, south to California, and east just into Alberta, and to Colorado and Texas. Under ideal conditions it may reach a height of 90 m. It is tolerant of a wide range of habitats, from relatively wet conditions along the coast (although here it is often found on well-drained, rocky soils) to dry conditions in the interior. The bark on old trees becomes thick and forms large, irregular square plates. Flattened grey-green needles are 2–3 cm long and may be produced in two ranks, horizontally or more or less evenly around the branches, giving a 'bottlebrush' effect. The smaller branchlets are usually pendulous, but the tips of main lateral branches are upturned. Over-wintering leaf buds are cinnamon brown and sharp-pointed. Cones are produced in great quantity and are borne on quite young trees. Mature cones are 5–10 cm long, pendulous, and reddish brown. They resemble those of no other local conifers because they have a 3-pronged bract below each rounded cone scale. Of native local conifers, Douglas Fir looks the greyest at a distance. *Pseudotsuga* means false-hemlock, although it looks only very superficially like our native hemlocks.

It is much too large a tree for most landscapes, but wild trees may been seen in abundance in Stanley Park, Pacific Spirit Regional Park, Queen Elizabeth Park, or any other natural area in the city. It is sometimes grown as an ornamental for a few years but very quickly becomes too large for the location and has to be removed or, too often, brutally pruned.

There are several cultivars, including the pendulous-branched form 'Pendula', in the BC Native Garden of UBC Botanical Garden and in the Western North American Section of VanDusen Botanical Garden.

Tsuga – Hemlocks

Often confused with spruces, the hemlocks also have rough twigs after the needles have fallen, but hemlock twigs are much more slender and less stiff than those of spruces. Hemlock needles are flat and blunt-tipped and, at least in our local species, tend to have needles of varying lengths on each twig, compared with the much more uniform length of spruce needles on any one twig. Several hemlocks have a distinctive drooping leader (top of the tree), especially in young trees. About ten species of hemlock are found in eastern North America, western North America, and eastern Asia, a distribution found in a number of other plant genera.

Tsuga canadensis (L.) Carr.
Canadian or Common Hemlock

This is the eastern North American counterpart of our Western Hemlock. It is a tall evergreen tree, growing to 30 m or more in cool, moist forests from Nova Scotia to Alabama. The needles do not appear quite as disarrayed as those of the Western Hemlock, and hold slightly more to two ranks along the twigs. A distinguishing feature of the needle arrangement is a row of short needles on the top of the twigs that are often turned so that the white bottoms of the needles are visible when viewed from above. The cones are up to 2 cm long, with each cone scale about as broad as it is long. They are slightly more rounded in outline and look a bit tidier than those of the Western Hemlock.

It is surprising that there seem to be so few of these in the city, although there may be others that have been overlooked because of their similarity to our common native Western Hemlock. The only specimens found are two trees in The Crescent.

Tsuga heterophylla (Raf.) Carr.
Western Hemlock

This important timber tree grows to 70 m tall and is found from Alaska to California and Idaho. Young trees have a distinct nodding leader (terminal growth). The needles are somewhat two-ranked, but always look in disarray, growing in varying lengths and sticking out in all directions from the twigs. The characteristic needles are reflected in the name *heterophylla*, meaning 'different leaves.' The needles are dark green above, with two white bands beneath. The very slender twigs feel rough after the needles drop.

Cones are about 2.5 cm long, with an irregular appearance and more elongated cone scales than those of other hemlocks. This is the only hemlock growing wild here on the coast at sea level, often found with Western Red Cedar and Douglas Fir. The dark green colour, very delicate branches, drooping tip, and small cones separate it from other similar coastal evergreens.

There are many wild trees in Stanley Park, Queen Elizabeth Park, and Pacific Spirit Regional Park. It is occasionally planted as a hedge or specimen plant in parks and gardens. Cultivated specimens may be seen on Robson St betweeen Gilford St and Denman St, and on Collingwood St between 30th Ave and 31st Ave.

Tsuga mertensiana (Bong.) Carr.
Mountain or Black Hemlock

Native of high elevations (1,000–2,500 m) from Alaska to northern California and western Montana, this is a very beautiful, but rarely cultivated native evergreen that tends to be slow-growing at lower elevations. It is not found in Vancouver, but is common around the ski areas on the north shore mountains. The branches have a 'bottlebrush' look from the needles being borne all around the twigs. The needles are slightly bluer than those of other hemlocks, and the cones are much longer (4–6 cm), looking more like a spruce cone than the cones of most other hemlocks.

It is not common here, but there are a few specimens, mostly small, around the city. There is a beautiful, perfectly sym-metrical specimen with limbs to the ground and bearing cones (growing with a White Fir) on the east side of MacKenzie St between 19th Ave and 20th Ave, a relatively large specimen in the Old Arboretum at UBC, smaller specimens in the Western North American Section at VanDusen Botanical Garden, and two in front of an apartment complex on the south side of 11th Ave just west of Granville St. •12

Taxaceae – Yew Family

Taxus – Yews

The yews and the junipers are the only 'cone-bearing' evergreens cultivated in our area that do not bear woody cones. Individual yew seeds are partially surrounded by a bright red, fleshy structure known as an aril, which looks much more like a berry than a typical conifer cone. The small, rounded male (pollen-bearing) cones are much more typical. Male cones are less than 1 cm across but are borne in such profusion on some trees that they give a yellow cast when the pollen is being shed, usually in February or March. Seeds are rarely produced on many yew trees, especially on our native Pacific Yew, but frequently on some dwarf clones of English Yew. The needles are flattened, forming two ranks along the branchlets. The very dark green colour is the best characteristic for separating the yews from other locally cultivated conifers, plus the fact that the needles are green, not grey beneath.

There are two extremely variable yews in cultivation, the English Yew (*Taxus baccata*) and Japanese Yew (*Taxus cuspidata*), plus hybrids between the two. These are most often seen as small, low-growing or upright shrubs used for hedges or as foundation plantings around homes, but these forms do become large trees eventually.

The yews are another good example of trees that are closely related, but isolated from one another geographically. In the wild, it is easy to put a name on any of the yews, but in cultivation the characteristics used to separate them are not as discernible and often inconsistent, especially when their origin is unknown.

Taxus baccata L.
English Yew

This European native is by far our most commonly cultivated yew. The dark green, glossy needles are 1–2 cm long and are paler green on the undersurface. There is no mistaking an English Yew in cultivation when it does bear the black seeds surrounded by the bright red, fleshy arils. Most specimens in the city were obviously planted as small multi-trunked shrubs but have now grown into trees up to 10 m tall, with several trunks that have often grown together at the base, forming an apparently single, irregular trunk.

There are large trees on the sw corner of Trimble St and 1st Ave, on the se corner of Oak St and 12th Ave, along the north side of Broadway between Vine St and Balsam St, on the NE corner of Oak St and Broadway (by the parking entrance to the BCAA Bldg), in The Crescent, on the south side of 5th Ave between Macdonald St and Bayswater St, and one on the NE corner of Stephens St and 3rd Ave that is taller than the two-and-a-half-storey house nearby.

Taxus brevifolia Nutt.
Pacific Yew

This is the only native yew on the Pacific Coast, and is found from Alaska to California and east to Idaho and Montana. It is not usually very common. Locally, most of the trees tend to be multi-trunked and rather shrubby, but the trees may be single-trunked and reach 10 m (rarely taller). The red, fleshy-coated seeds are rarely produced on local trees. The characteristic often used to distinguish Pacific Yew from English Yew is that the winter bud scales are keeled, but this is difficult to see and separate from the rounder bud scales of the cultivated English Yew. Older specimens of Pacific Yew have a distinct cinnamon red bark, much more colourful than that of the English Yew, and this seems to be the best characteristic for separating the two locally. One can usually assume that if the tree is in cultivation in our area, it is likely to be English Yew, not Pacific Yew. The trees have attracted a great deal of attention recently when it was found that the bark contains the chemical taxol, used in the control of some forms of cancer.

There are some trees scattered around Pacific Spirit Regional Park and Stanley Park. There is a nice, relatively large specimen that can be seen easily at Brockton Point in Stanley Park, one on the north side of the Pitch & Putt Golf Course in Stanley Park, and several young ones in the BC Native Garden at UBC Botanical Garden. •13

Taxodiaceae – Taxodium Family

Cryptomeria japonica (L.f.) D. Don
Japanese Cedar

This evergreen tree grows to 35 m tall in nature in China and Japan, where it is a major timber tree. The short needle-like, slightly curved leaves lie nearly flat against the branchlets. The cones are rounded, about 2–3 cm across, and covered with soft prickles.

Probably the largest tree in Vancouver is SE of the Stanley Park Dining Pavilion (toward the Zoo). There are large ones on the east side of Marguerite St north of 45th Ave (hanging over the sidewalk), in the Old Arboretum at UBC, two in the middle of a long row of catalpas between Buchanan Bldg and the Clock Tower (between Main Mall and East Mall) at UBC, a long row on both sides of Thunderbird Blvd from West Mall (by the parking lots on the south side of UBC campus), and a relatively large one west of the monument in Memorial South Park off 41st Ave.

'Cristata' – This cultivar looks much like the typical wild form except that the ends of some of the branches are flattened in a fan shape or cockscomb-like (fasciated). There are two on the north side of the Faculty Club at UBC, a large one in a yard on the east side of Marguerite St just north of 45th Ave (among typical wild forms), one between the Pitch & Putt Golf Course and the tennis courts in Stanley Park and another east of the tennis courts.

'Elegans,' Plume Cryptomeria – This cultivar, which is usually seen as a shrub, has longer needles that are very feathery and soft to the touch. It eventually reaches small-tree proportions. The needles are bright green in summer and turn a distinctive purple to reddish brown in autumn and winter. There are large specimens of this variety on the NW corner of 14th Ave and Waterloo St, on the western side of Bobolink Ave at Muirfield Dr, on the north side of 29th Ave between Camosun St and Crown St, and several on the NW corner of Heather St and 37th Ave. A similar cultivar 'Elegans Aurea' remains green over winter. There are small specimens of this cultivar in the Winter Garden at UBC Botanical Garden, growing with 'Elegans' for comparison.

There are many other unusual cultivated forms with various colours, textures, sizes, and shapes, but only a few of these become trees. There are a number of different cultivars west of the Heather Garden in VanDusen Botanical Garden, and in UBC Botanical Garden.

Cunninghamia lanceolata (Lamb.) Hook.
Chinese Fir or China Fir

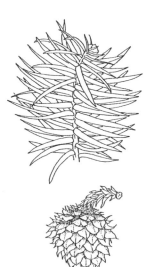

This soft-looking evergreen has flat, sickle-shaped needles (4–6 cm long) that are actually very prickly to the touch. The typical wild form found in China is bright green, but there are also grey-green forms that are more popular in cultivation. Although it may grow to be 30 m tall in the wild, most of the ones in the city are relatively young and not very large. Cones, 3–5 cm long, are nearly round, with sharp points to their scales.

Among the green forms are specimens in the Old Arboretum on the UBC campus, on the south side of 13th Ave at Sasamat St, a large one on the NW corner of Chancellor Blvd and Newton Cres (towering above a hedge), and an old specimen in the median of Angus Dr between Hosmer Ave and Granville St.

'Glauca' – The very grey form. There are specimens SW of the Main Library at UBC, on the left side of the entrance to Hycroft on McRae St, and a number of small trees in the Sino-Himalayan Garden at VanDusen Botanical Garden.

Metasequoia glyptostroboides Hu & Cheng
Dawn Redwood

This fast-growing, deciduous conifer is native to China and was known only from fossils before live trees were discovered in 1941. Seeds were sent to the Arnold Arboretum in Boston a few years later, and trees were distributed from there to many botanical gardens and parks around the world. Obviously, there are no really old trees in cultivation, so it is not known just how large they will grow, but they do grow very quickly and even young trees develop a distinctive buttressed trunk. Based on the fossils, the tree originally was assumed to be an evergreen. Soft, pale green leaves are borne on opposite pairs of deciduous branchlets, turning bronze in the autumn. Those of the similar Bald Cypress (*Taxodium distichum*) are borne alternately along the branches. Fairly young trees produce very attractive and distinctive oval cones, 2–3 cm long.

There is a street planting on Arbutus St from 32nd Ave north for two blocks, and a row planted about 1964 on Kerr St between

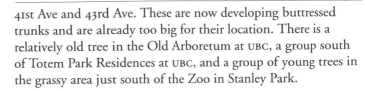

41st Ave and 43rd Ave. These are now developing buttressed trunks and are already too big for their location. There is a relatively old tree in the Old Arboretum at UBC, a group south of Totem Park Residences at UBC, and a group of young trees in the grassy area just south of the Zoo in Stanley Park.

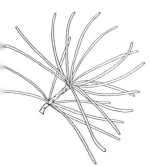

Sciadopitys verticillata (Thunb.) Siebold & Zucc.
Umbrella Pine

This beautiful evergreen tree is a native of central Japan and is often cultivated in botanical gardens and parks in temperate climates, although it is not generally known to the public. The long, bright green needles, 8–12 cm long, are produced in whorls like the stays of an umbrella, thus the common name. The needles are dark green above, with two white stomatal bands beneath, and have a very hard, plastic feel. The cones, about the same length as the needles, are relatively soft and pliable when mature. They are not usually produced until the trees are relatively old.

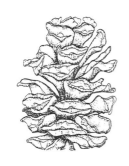

The best specimen in the city is an unusually dense one on the grounds of St. George's School on the north side of 29th Ave between Wallace St and Highbury St. Other large specimens include ones in the Old Arboretum and just south of the Main Library at UBC, two large specimens on Pine Cres near Pine St and at Hosmer Ave, one with a multiple top on the SE corner of Matthews Ave and Selkirk St, one on the SE corner of Cartier St and 47th Ave, and a very dense, smaller tree on the north side of King Edward Ave west of Alexandra St.

Sequoia sempervirens (D. Don) Endl.
Coast Redwood

This is one of the tallest trees in the world (possibly matched only by some *Eucalyptus* species), growing to as much as 100 m tall. It is native to the fog belt of coastal California and just into southern Oregon. The trunks are slender and less massive than the Giant Sequoia, and the trees do not live to be quite so old, only to about 1,500 years. There are two types of needles: small and scale-like on the vigorous terminal shoots; and flat dark green needles, 1.5–2.5 cm long, in flattened sprays on most of the branches. The red-brown cones are 2–3.5 cm long, about half the size of those of the Giant Sequoia. It does not grow as well this far north as does the Giant Sequoia.

Among the larger ones around the city are specimens on the grounds of Point Grey Secondary School at East Blvd and 37th

Ave, on the west side of Macdonald St between 13th Ave and 14th Ave, on the north side of the Fish House Restaurant south of the Pitch & Putt Golf Course in Stanley Park, and several among the bamboo planting at VanDusen Botanical Garden.

Sequoiadendron giganteum (Lindl.) Buchh.
Giant Sequoia or Bigtree

This evergreen tree is not quite as tall in nature as the Coast Redwood, but it does reach nearly 100 m and is much more massive. In fact, it is the most massive of all trees. Some individuals are thought to be 3,500 years old and are among the oldest living things. The old trees have strongly buttressed trunks and often have no limbs for up to 50 m above ground. Young trees grow very quickly and keep their branches to the ground for many years, but quickly develop the characteristic buttressed trunk. The awl-like, grey-green needles are 3–6 mm long, or longer on vigorous young growth. The oblong female cones, about 5–8 cm long, are hard and heavy. They are about twice the size of the cones of the Coastal Redwood, but smaller than would be expected on such a large tree. In nature Bigtrees are now found only in small groves on the western slope of the Sierra Nevada of California.

It is a fairly commonly cultivated tree, but is often placed where it doesn't have enough room to grow and has to be removed after a few years. Anyone thinking about planting a Bigtree should look at some of the following relatively young but massive trees. All are mere babies, having been planted in the last 60 to 70 years. There are two on the grounds of Point Grey Secondary School at East Blvd and 37th Ave, one in front of the Main Library at UBC, one on the SE corner of 49th Ave and Granville St, one on the north side of 41st Ave at Vine St, a row in the median of Cambie St between King Edward Ave and 29th Ave, and two large ones on the west side of Marguerite St just north of 49th Ave. There is a beautiful grove of young trees at VanDusen Botanical Garden that were planted in 1973 and which have become surprisingly large in a short time. •14, 85

'Pendula,' Weeping Giant Sequoia – This is a very slender, erect but sinuous, tree form with all the side branches hanging close against the trunk. It was found as a genetic mutation on a tree growing in France and is now grown in collections around the world for its interesting growth habit. The best examples of this tree are in the Alpine Garden and the Conifer Collection at VanDusen Botanical Garden. There are others along the entrance to Shaughnessy Golf and Country Club off SW Marine

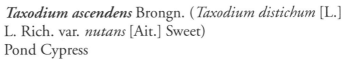

Dr near 41st Ave, a group in a garden on the north side of 47th Ave between Oak St and Montgomery St, one in the SE corner of the Quarry Garden in Queen Elizabeth Park, and one in a roof garden at the corner of Alberni St and Thurlow St. •15

Taxodium ascendens Brongn. (*Taxodium distichum* [L.] L. Rich. var. *nutans* [Ait.] Sweet)
Pond Cypress

This deciduous conifer is considered by some authorities to be merely a variety of the *Taxodium distichum*, the more common Bald Cypress. But others consider it to be distinct, as it is smaller and has thread-like branchlets rather than the more feathery ones of Bald Cypress.

The only two known in the city are a large specimen in the Old Arboretum at UBC, and a younger one in the Eastern North American Section at VanDusen Botanical Garden.

Taxodium distichum (L.) L. Rich.
Bald Cypress

One of the few deciduous cone-bearing trees native to North America, this is a familiar sight in the southeastern United States where it grows partially submerged in swamps. Old trees often have buttressed trunks and woody projections called 'knees' that grow up from the roots and extend out of the water. These probably help with air intake for the waterlogged roots. The pale green, soft, needle-like leaves are borne along alternately arranged branchlets. The whole small branchlet turns coppery-coloured before shedding in the autumn. This tree and the Dawn Redwood (*Metasequoia glyptostroboides*) are very similar in appearance, but the alternate branches distinguish the Bald Cypress. Cones are oval or egg-shaped, 2–3 cm long, and break apart as they ripen.

This is not a very common tree in Vancouver. There are four in Queen Elizabeth Park (one in the centre of the Quarry Garden and three on the north side just south of 29th Ave), a single specimen on the hillside in Almond Park at Dunbar St and 13th Ave, one on the NE corner of 6th Ave and Dunbar St, two in the median of Angus Dr between Hosmer Ave and Granville St, and several young specimens growing on the edge of Cypress Pond in the Eastern North American Section in VanDusen Botanical Garden which are already beginning to produce knees.

Angiosperms:
Flowering Plants

Aceraceae – Maple Family

Acer – Maples

The maples are familiar, usually deciduous, trees that always have leaves borne in opposite pairs along the twigs. The leaves are most often palmately-lobed and toothed, but they may be simple and unlobed or compound. A combination of the opposite leaves and pairs of winged fruits, technically known as samaras and commonly known as keys, are the two distinguishing characteristics that separate maples from all our other temperate trees.

There are three species of native maples in British Columbia, two of which are common in the city: the Big-Leaf Maple (*Acer macrophyllum*) and Vine Maple (*Acer circinatum*). A third, Douglas Maple (*Acer glabrum*), is found on the Gulf Islands and is common in the interior of the province, but it is rare locally. There are many introduced maples planted in the city as street trees, shade trees, or garden ornamentals.

In addition to the relatively common maples dealt with in this book, there are many other rare species and cultivars which may be represented by only one or two trees in the Maple Collection and Sino-Himalayan Garden at VanDusen Botanical Garden and in UBC Botanical Garden, especially in the Asian Garden. The only other member of the maple family, *Dipteronia sinensis,* has pinnately-compound leaves and looks very unlike the maples. It has rounded elm-like samaras attached in pairs. There is a young one in the Asian Garden (near the Administration Bldg) at UBC Botanical Garden.

Acer buergeranum Miq.
Trident Maple

This Chinese maple grows to about 15 m tall and has relatively thick leaves with 3 distinct lobes. It is not seen often in cultivation in our area. The leaves are often slightly cup-shaped rather than flat, and are shiny dark green above and glaucous beneath, turning orange or red in autumn. The wings of the samaras generally overlap at their tips.

This tree is rare here. There is one on the berm along Stadium Rd off sw Marine Dr outside the UBC Botanical Garden, and in VanDusen Botanical Garden there are two by the lake in the Maple Collection, as well as a number in the Sino-Himalayan Garden.

Acer keys

1 A. buergeranum
2 A. campestre
3 A. capillipes
4 A. cappadocicum
5 A. circinatum
6 A. cissifolium
7 A. davidii
8 A. ginnala
9 A. griseum
10 A. glabrum
11 A. japonicum
12 A. macrophyllum
13 A. negundo
14 A. nikoense
15 A. palmatum
16 A. platanoides
17 A. pseudoplatanus
18 A. rubrum
19 A. rufinerve
20 A. saccharinum
21 A. saccharum
22 A. shirasawanum
23 A. tataricum

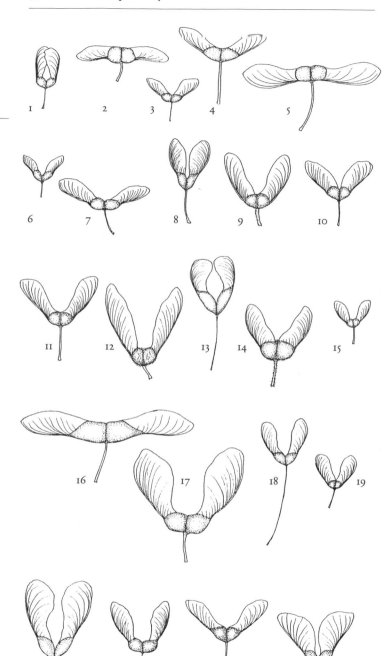

Acer campestre L.
Hedge Maple or Field Maple

This attractive, commonly cultivated maple from Europe grows to about 10 m tall or occasionally taller. The small 3–5-lobed leaves (to 10 cm wide) with very rounded lobes distinguish it from all our other local maples. The keys have wide spreading wings. Autumn colour is often bright golden yellow. It is very frequently planted as a street and park tree in Vancouver.

Probably the largest tree (with several other large ones nearby) is in Stanley Park between the perennial beds and the causeway, just NE of the overpass walkway. There are good street plantings along Wallace St from 20th Ave to 22nd Ave, along 16th Ave from Oak St to Cambie St (alternating with hawthorns), on the east side of Ontario St north of 33rd Ave, and a number of trees around Balaclava Park (between 29th Ave and 31st Ave, and Carnarvon St and Blenheim St).

Acer capillipes Maxim.
Red Snake-Bark Maple

This is the first of several maples in a group known as the snake-bark maples, all of which have beautiful, slick, pale green bark with narrow vertical stripes of white, especially on the twigs, young limbs, and trunks of young trees. They also tend to have relatively thin, soft leaves. The Red Snake-Bark Maple is a native of Japan, growing to 10–12 m tall, with good red autumn colour. The leaves usually have three, or sometimes five, major lobes, with small teeth. The yellow-green flowers are borne in long drooping chains, followed by small keys. It is very similar to *Acer rufinerve,* but lacks the tufts of reddish hairs on the veins underneath the leaves.

There are specimens on the north and south sides of the Stanley Park Pitch & Putt Golf Course, and in the Asian Garden at UBC Botanical Garden.

Acer cappadocicum Gled.
Cappadocian Maple

This attractive maple, growing to about 20 m tall, is native to areas extending from the Caucasus, Turkey, and Iran to western China. The thin 5–7-lobed leaves superficially resemble a small London Plane Tree leaf (*Platanus × acerifolia*). The young leaves in spring are coppery or purplish green, followed closely by clusters of green flowers in rounded heads, much like those of the Norway Maple. Flowers open about mid-April in our area.

The autumn colour is usually a good, bright gold. It makes a very attractive, round topped street tree.

There is a street planting on both sides of 12th Ave between Fir St and Maple St, alternating with green- and purple-leaved forms of Norway Maple (the leaves of the Norway Maple are much larger); a row in the median of the walkway into Stanley Park (by the causeway); a large tree behind the Education Bldg along University Blvd, just west of Main Mall on the UBC campus; and several in the Maple Collection at VanDusen Botanical Garden.

'Aureum' – The golden-leaved cultivar. There is a nice specimen against an apartment block on the north side of York St just west of Arbutus St and a relatively large one in the Maple Collection at VanDusen Botanical Garden.

Acer circinatum Pursh
Vine Maple

The name Vine Maple is a misnomer as the tree is not a vine at all, but neither is it a large tree. It usually becomes a clumped, multi-trunked small tree growing to about 10 m tall. It is common in the edges of forests in the southwestern part of British Columbia. The relatively small, broad leaves have 7–9 lobes and they are pale green in summer, turning rich yellow, orange or red in autumn, making it one of our most colourful deciduous trees. The dark red and white flowers are moderately showy against the pale green new leaves in spring. The flowers are followed by samaras with wide spreading wings. It is closely related, and very similar in leaf size and shape, to the typical wild form of the Japanese Full-Moon Maple (*Acer japonicum*). Vine Maple deserves to be planted as an ornamental in our gardens more than it is. It is rarely used as a street tree and not often seen as a single-trunked tree.

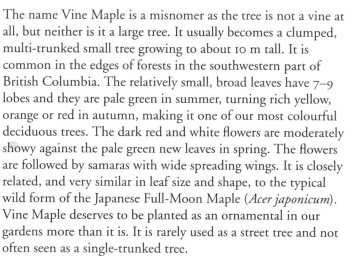

It is a very common native tree in the edges of forests and along trails around the city, especially in Stanley Park, Queen Elizabeth Park, Pacific Spirit Regional Park, and the University Endowment Lands. There is a very typical group along the east side of Trimble St at Hadden Ave at Spanish Banks; a group in Kitsilano Beach Park just west of the Maritime Museum; large ones on the west side of Shaughnessy St just north of 67th Ave; and street plantings of single-trunked trees along Hudson St between 32nd Ave and 33rd Ave, and along Marguerite St between 37th Ave and 41st Ave. •16

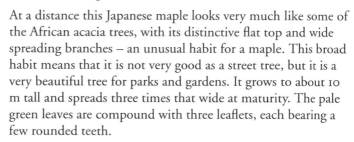

Acer cissifolium (Siebold & Zucc.) C. Koch
Vine-Leaf Maple

At a distance this Japanese maple looks very much like some of the African acacia trees, with its distinctive flat top and wide spreading branches – an unusual habit for a maple. This broad habit means that it is not very good as a street tree, but it is a very beautiful tree for parks and gardens. It grows to about 10 m tall and spreads three times that wide at maturity. The pale green leaves are compound with three leaflets, each bearing a few rounded teeth.

Slender, pendulous, catkin-like flower clusters of yellow-green are produced just as the leaves emerge in late April or May.

There is a large individual on the UBC campus at the corner of West Mall and University Blvd and three smaller specimens in front of the West End Community Centre on Denman St. The largest street plantings are around MacDonald Park (between Sophia St and Prince Edward St, and 44th Ave and 45th Ave), where they alternate with the similar but more common and more upright-growing *Acer negundo,* which has leaves with 3–5 leaflets. Probably the largest specimen in the city was a very beautiful one by the parking lot of Vancouver General Hospital, but it was cut down in early 1989 for expansion of the hospital.

Acer davidii Franch.
Père David's Maple

Probably the most beautiful of the snake-bark maples, Père David's Maple is another of this group of similar ornamental maples from China. A small tree, growing 10–15 m tall, it deserves to be grown more often than it is. Twigs and the young smooth bark have bright vertical stripes of white and pale green, contrasting nicely with the very dark green leaves. These are ovate to ovate-oblong and are toothed, but usually unlobed, or with 2 obscure lobes at the base, a characteristic that will separate it from most of the other snake-bark maples which have lobed leaves. The small keys are produced in long, drooping chains and are prominent in late summer and autumn.

There is a small tree in one of the beds on the western side of the perennial beds in Stanley Park (near the causeway and west of the Rose Garden), two in the Winter Garden and several in the Asian Garden at UBC Botanical Garden, and some along the Rhododendron Walkway and in the Sino-Himalayan Garden at VanDusen Botanical Garden. •17

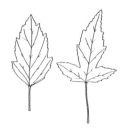

Acer ginnala Maxim.
Amur Maple

This small tree grows to about 7 m tall. A native of Manchuria and Japan, it has glossy, rather un-maple-like leaves, either with three large lobes or simply oblong with irregular teeth around the margins. The wings of the samaras are nearly parallel and the samaras are usually bright green when they first develop, later turning orange or red. Young growth in spring is distinctively pale green, turning dark green for summer and sometimes a good orange or red in autumn.

There are a few street plantings in the city, including along 30th Ave from Gladstone St to Baldwin St, and on 14th Ave from Windsor St to Prince Albert St. Two of the largest specimens are on the SE corner of 15th Ave and Sasamat St. There are several old, multi-trunked specimens at the corners of the Math Bldg on Main Mall and on the SW corner of the East Mall Classroom Block on the UBC campus, and several large specimens in front of Vancouver Technical Secondary School on the south side of Broadway between Penticton St and Slocan St.

Acer glabrum Torr. var. *douglasii* (Hook.) Dipp.
Douglas Maple

This small tree is native to western North America from southern British Columbia to California. It is rare in Vancouver but more common to abundant on rocky areas of the Gulf Islands, southern Vancouver Island, and especially east of the Coast/Cascade Mountains. The dark green leaves are grey beneath and 3–5-lobed, occasionally so deeply lobed as to form 3 distinct leaflets. The samara wings are almost parallel.

The only specimens seen in the city are in the Old Arboretum on the UBC campus, in the BC Native Garden of the UBC Botanical Garden, and in the Western North American Section of VanDusen Botanical Garden.

Acer griseum (Franch.) Pax
Paperbark Maple

The distinctive and unique feature of this lovely maple is its bright cinnamon brown bark that peels away in thin papery flakes, thus the common name. It is a native of China. The compound leaves have 3 leaflets, large blunt teeth, and are rather pale green above and soft grey-green beneath. Leaves become a muted orange-red in the autumn. Keys have large seeds compared to the relatively short wings, and are also soft-pubescent. Unfortunately, the keys are usually empty, making propagation of this maple difficult.

There are specimens east of the tennis courts in Stanley Park, in the Alpine Garden and Asian Garden at UBC Botanical Garden, and in the Sino-Himalayan Garden at VanDusen Botanical Garden. •18, 19

Acer japonicum Thunb.
Full-Moon Maple

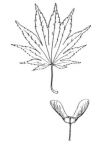

This very attractive Japanese maple is much less common in gardens than another Japanese maple, *Acer palmatum*. It often remains a small, bushy tree, but may grow to about 7 m tall. The typical form is rarely cultivated but looks so similar to our native Vine Maple (*Acer circinatum*) that the two are not easily separated. However, the form usually cultivated here is very distinctive and easily recognizable. There are wild forms and several other rare cultivars grown locally only in the Asian Garden of UBC Botanical Garden.

'Aconitifolium' – A relatively common cultivar with very deeply dissected, large leaves to 15 cm wide, with overlapping lobes. The leaves turn a vivid red in the autumn. There is a slender specimen about 3 m tall in the lawn area just north of the tennis courts off Park Dr in Stanley Park, one against an apartment building on the east side of Balsam St between 2nd Ave and 3rd Ave, a large one on the east side of the larger quarry in Queen Elizabeth Park, one in the Alpine Garden and several in the Asian Garden at UBC Botanical Garden, and several in the Maple Collection at VanDusen Botanical Garden.

Acer macrophyllum Pursh
Big-Leaf Maple

When in leaf, the Big-Leaf Maple cannot be confused with any other deciduous tree. The trees are the most massive of native Canadian maples, and they have the largest leaves of any native tree in Canada as well as the world's largest maple leaf! It is common along the coast from southeastern Alaska to California. Leaves, 30 cm or more wide, are deeply lobed, but are variable in size and shape from tree to tree or limb to limb. Young, vigorous growth usually has the largest leaves. Occasionally, the leaves turn a good golden yellow in autumn, but in other years the leaves only become dull green before dropping. In spring, usually in March to April, the large, pendulous clusters of yellow-green flowers are among the showiest of maple flowers. The seed portion of the samara is covered with sharp, bristly hairs that can be painful when they contact tender skin.

This is a common native tree in Stanley Park and in other wooded areas of the city. There are many nice old specimens near the concession stand at Third Beach in Stanley Park and around the statue of Lord Stanley near the park entrance, three very large trees on the north side of a small park between Willow St and Laurel St and 26th Ave and 27th Ave, large individuals on the sw corner of Ash St and 16th Ave and on the north side of 49th Ave at St. George St, and a street planting of relatively large trees on the east side of Maple St between 39th Ave and 41st Ave. •20

Acer negundo L.
Box-Elder, Manitoba Maple, or Ash-Leaved Maple

This is our only native Canadian maple with compound leaves, with 3–7 irregularly toothed or lobed leaflets, which give it a very atypical look for a maple. However, the slender keys with nearly parallel wings are typical of all maples. The long chains of keys hanging on bare winter branches are a distinctive feature of female trees of this maple in winter. It is native to central and eastern North America but is often cultivated and escaped from cultivation elsewhere, as in northeastern British Columbia.

It is fairly common in cultivation in Vancouver. There are specimens at Hudson St and 52nd Ave, a street planting on Wiltshire St from 52nd Ave to 54th Ave, a row of relatively large specimens on 10th Ave from Willow St to Laurel St (inter-planted with elms), several around MacDonald Park (between 44th and 45th Ave and Prince Edward St and Sophia St), and four trees on Pendrell St between Denman St and Gilford St.

'Variegatum' – This cultivar, with leaves edged in white, is a very attractive deciduous tree. Even the keys are variegated, usually with the seed being green and the wing white. Limbs with all green leaves often appear scattered throughout the trees. There is a large specimen on the grounds of Vancouver City Hall along 12th Ave at Cambie St, a large one on the north side of the road to Bloedel Conservatory in Queen Elizabeth Park, one on the nw corner of 4th Ave and Trutch St, one on the se corner of 58th Ave and Cambie St, and a street planting on 36th Ave from East Blvd to Pine Cres.

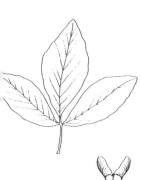

Acer nikoense Maxim. (*Acer maximowiczianum* Miq.)
Nikko Maple

Another compound-leaved maple, the Nikko Maple has leaves with three leaflets and closely resembles the Paperbark Maple (*Acer griseum*) except that the leaves of the Nikko Maple are a bit larger (6–10 cm long), less toothed, and less grey beneath. The most distinctive difference is that the bark does not peel as in the Paperbark Maple. It is a small tree reaching 15–20 m tall in its native China and Japan, giving good red and orange colours in the autumn. It may be listed in some references as *Acer maximowiczianum,* but the correct name for the tree has been shown recently to be *Acer nikoense,* which is at least easier to pronounce. The tree is very rarely cultivated outside botanical gardens and arboreta.

There is one tree on the NW corner of the Stanley Park Pitch & Putt Golf Course, one on the NE side of Queen Elizabeth Park, and one in the Asian Garden at UBC Botanical Garden.

Acer palmatum Thunb.
Japanese Maple

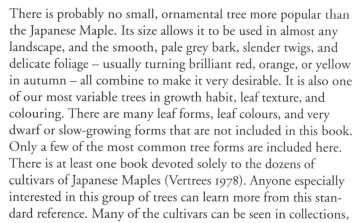

There is probably no small, ornamental tree more popular than the Japanese Maple. Its size allows it to be used in almost any landscape, and the smooth, pale grey bark, slender twigs, and delicate foliage – usually turning brilliant red, orange, or yellow in autumn – all combine to make it very desirable. It is also one of our most variable trees in growth habit, leaf texture, and colouring. There are many leaf forms, leaf colours, and very dwarf or slow-growing forms that are not included in this book. Only a few of the most common tree forms are included here. There is at least one book devoted solely to the dozens of cultivars of Japanese Maples (Vertrees 1978). Anyone especially interested in this group of trees can learn more from this standard reference. Many of the cultivars can be seen in collections, as in VanDusen Botanical Garden and UBC Botanical Garden.

The typical wild forms are usually multi-trunked trees, growing to about 8–10 m tall, with small, palmately-lobed, green leaves (5–10 cm long and about as wide) with 5–9 long pointed lobes and small teeth along the edges. The most popular forms in cultivation are purple-leaved forms and weeping, dissected-leaved forms with either green or purple leaves. There is a naturally occurring variety known as *heptalobum,* with leaves at the larger end of the scale which are usually 7-lobed. •21

Wild Green-Leaved Forms (var. *palmatum* **and var.** *heptalobum*) – There are many in the city in most parks and many private gardens. There are a number of trees of variable sizes in the median strip of 16th Ave from Blenheim St to Collingwood St, several large ones NW of the Stanley Park Dining Pavilion and around Robson Square, a number of large ones in Nitobe Garden and the Asian Garden at UBC Botanical Garden, a large one in the garden area NW of Main Library at UBC, and many in the Maple Collection around one of the lakes at VanDusen Botanical Garden. It is very unusual to see it as a street tree, but there is a nice row of large specimens on the north side of 57th Ave from Rosemont Dr to Vivian Dr.

'Atropurpureum' – This cultivar is like the typical wild, tree form, but the leaves are larger and very dark purple. It is common here. There is a large one in the garden area NW of the Main Library at UBC, a large one between the Stanley Park Dining Pavilion and the Zoo, a number in and around the Asian Section of the Alpine Garden at UBC Botanical Garden, several in the Maple Collection at VanDusen Botanical Garden, several at Lumberman's Arch in Stanley Park, and several in Almond Park (12th Ave and Dunbar St).

'Dissectum' – A popular cultivar usually seen as a small fountain-shaped tree with very pendulous branches that completely hide the trunk in summer, making it look like a rounded shrub. Only in winter does the trunk show. The green leaves are very dissected and frilly, giving a lacey appearance to the tree. There are several specimens in the Asian Section of the Alpine Garden at UBC Botanical Garden, in the Maple Collection at VanDusen Botanical Garden, and in and around the Quarry Garden at Queen Elizabeth Park.

'Dissectum Atropurpureum' – A very popular cultivar, it is nearly identical to the above form except that the leaves are dark purple. There are a number of variants on this form with slightly differently shaped or coloured leaves. Good specimens around the city include ones north of the Rose Garden and perennial beds and around the Dining Pavilion in Stanley Park, in the Maple Collection at VanDusen Botanical Garden, in and around the Quarry Garden at Queen Elizabeth Park, and on the south side of 10th Ave between Crown St and Camosun St.

'Linearilobium' – This cultivar forms a small tree. It has green leaves dissected to the base, with 5–7 very long, slender lobes. There is one in front of the Parks Board Office on Beach Dr, one by the lake in the Maple Collection at VanDusen Botanical Garden, and a large one recently planted in front of the Shop-in-the-Garden at the entrance to UBC Botanical Garden.

'Ribesifolium' – A stiff, upright-growing cultivar with irregularly round-lobed leaves. There is a nice specimen in the smaller Quarry Garden at Queen Elizabeth Park, and two by the lake in the Maple Collection at VanDusen Botanical Garden.

'Sangokaku,' Coral Bark Maple – This pale green cultivar looks relatively nondescript in summer but comes into its own in winter and early spring when the bright red twigs are very showy. There is a group of small ones on the west side of the new wing of Vancouver General Hospital along Oak St just south of 10th Ave, individuals on the north side of 27th Ave between Cambie St and Ash St and on the south side of the Stanley Park Pitch & Putt Golf Course, and one in the Maple Collection at VanDusen Botanical Garden.

Acer platanoides L.
Norway Maple

The round-topped crown of Norway Maple is especially distinctive in old trees growing on their own. It is native to Europe and western Asia, grows to 20 m or more tall, and is very commonly cultivated as a city street or garden tree in North America and Europe. It is seen as the typical green-leaved form or in many leaf colours and growth habits. The leaves are 5-lobed with an additional pair of points on each lobe. They are more or less the same shade of green on both sides, a characteristic that can be used to distinguish its leaves from those of some other similar maples, especially Sycamore Maple (*Acer pseudoplatanus*). Long petioles, usually longer than the leaf blade, are also distinctive. Yellow-green flowers are produced in mid- to late March, just as the new leaves are emerging. These rounded flower clusters are moderately showy when produced in profusion, as they are on some trees. They are especially showy against the purple leaves of some cultivars. The sap is milky, especially in young, actively growing twigs. Keys are very distinctive, being relative large and flat, with the wings spreading nearly horizontally. The milky sap and shape of keys will readily distinguish Norway Maple from the similar Sugar Maple (*Acer saccharum*), which has clear sap and keys with nearly parallel wings.

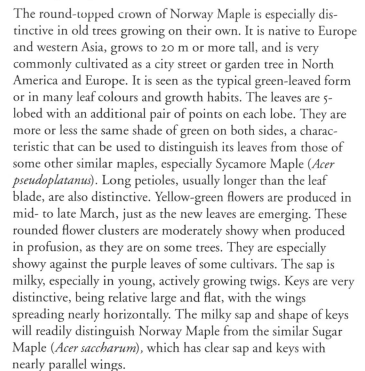

Norway Maple is a very common street tree here, often re-seeding and growing in vacant lots and sometimes seemingly native, as it is in Pacific Spirit Park and the University Endowment Lands. The many street plantings of green forms include those on 15th Ave from Blanca St to Crown St, on 13th Ave from Waterloo St to Trutch St and from Carnarvon St to Macdonald St, on 3rd Ave from Vine St to Balsam St, and on East Blvd from 42nd Ave to 49th Ave.

'Columnare' – There are a number of modern, named cultivars with a broad or narrow columnar shape. They are all lumped here under the cultivar 'Columnare,' although there are probably several different forms in the city. There is one in the Maple Collection at VanDusen Botanical Garden, two in the park at Argyle St and 43rd Ave, one on the south side of 16th Ave at Stephens St (among other typical forms of the species), and one NW of the Main Library at UBC (with typical forms). An individual columnar tree with purple leaves on the hill east of the restaurant in Queen Elizabeth Park seems to be an unnamed form.

'Crimson King'

'Crimson King' ('Schwedleri Nigra') – A better, darker selection of 'Schwedleri' with very dark, purplish black leaves that hold their colour all season. It has become more popular in recent years than 'Schwedleri.' There is one on the north side of Wolfe St at 16th Ave, one in the traffic circle in front of Bloedel Conservatory in Queen Elizabeth Park, one along the road to the Conservatory on the east side of the hill (with a large variegated *Acer negundo*), one on the NE corner of University Blvd and Western Parkway, and a row along West Blvd from 41st Ave to 49th Ave (alternating with Pin Oaks).

'Crispum'

'Crispum' ('Laciniatum'), Eagle's Claw Maple – An odd-looking cultivar with twisted, usually cup-shaped leaves. It is certainly more of a curiosity than attractive. The only one seen in the city is on the NW corner of the Pitch & Putt Golf Course in Stanley Park.

'Dissectum'

'Dissectum' – A cultivar with very deeply lobed leaves. There are two on the NE corner of 7th Ave and Prince Albert St.

'Globosa' – This tree has a very dense rounded head on a straight trunk, so that it looks like a large green lollipop. It is fairly common around parks in the city and as a street planting. There are several in Queen Elizabeth Park near the top of the hill in the median strip opposite the parking area for the

Seasons in the Park Restaurant, along 22nd Ave from Commercial Dr to Victoria Dr, on the west side of Ross St between 41st Ave and 43nd Ave, along the north side of 27th Ave between Crown St and Wallace St, and in Balaclava Park along the south side of 29th Ave between Carnarvon St and Balaclava St.

'Harlequin' ('Drummondii') – A popular and distinctive cultivar with creamy white margins around the leaves. The edges become yellow-green by late summer and may burn easily under hot, windy conditions. There are five young trees on the SE corner of Clark Dr and Broadway in front of Queen Alexandra Elementary School; five on the SW corner of 6th Ave and Prince Albert St; two on the east side of Prince Rupert St at 44th Ave; and one each on the north side of 33rd Ave at Yew St, the north side of College Highroad between Adelaide Rd and Chancellor Blvd, on the west side of Beechwood St between 53rd Ave and 54th Ave, and in the Maple Collection at VanDusen Botanical Garden.

'Schwedleri' – This was for years the standard dark-leaved cultivar of Norway Maple. It has now been largely replaced by better forms, such as 'Crimson King.' Leaves are coppery purple in spring and turn dull coppery for summer, similar to the colour of copper beeches and purple-leaved plums. It is very similar to 'Crimson King,' but not as dark, and it pales during the summer, while 'Crimson King' holds the darker colour. There are street plantings along 12th Ave from Maple St to Fir St (alternating with *Acer cappodocicum*), along King Edward Ave between Wallace St and Dunbar St, along 16th Ave from Wallace St to Dunbar St, along Crown St from 12th Ave to 16th Ave, three trees on the east side of Oak St at 14th Ave, and one in Stanley Park just west of the Pitch & Putt practice greens off Beach Ave. One of the largest trees is in the courtyard behind the Chemistry Bldg at UBC.

Acer pseudoplatanus L.
Sycamore Maple

This is a large maple native to Europe and western Asia with 5-lobed leaves and toothed margins. Leaves are slightly smaller and narrower than those of the Norway Maple, the lobes are rounder, and the keys are smaller and their wings less spreading. Leaves are often coppery when young, becoming dark green above and usually purple, white, or grey-green beneath with age. The greenish yellow flowers are produced in long hanging

chains rather than the rounded clusters of the Norway Maple, and are produced about 3–4 weeks later than those of the Norway. Bark on old trees develops into irregular plates of various shades of browns, more like that of the sycamores (*Platanus*) than other maples. Indeed, the tree is known as Sycamore in many places.

This tree is not quite as commonly cultivated here as the Norway Maple, but there are a number of street plantings and it is fairly common in parks and in public and private gardens. It reseeds and can be found naturalized in Stanley Park, Pacific Spirit Regional Park, the University Endowment Lands, and around Point Grey. Street plantings are found along the east side of Maple St between 13th Ave and 14th Ave and along 33rd Ave between Arbutus St and Larch St and between Carnarvon St and Blenheim St; there are also a number of large specimens in Tatlow Park between Point Grey Rd and 3rd Ave west of Macdonald St.

'Atropurpureum' ('**Spaethii**') – A selection with leaves darker green above and very purple beneath. Many trees exhibit some of these characters, but this form is the most distinctively purple beneath. There is a row on the east side of Oak St between 10th Ave and 11th Ave, along Heather St from Broadway to 11th Ave, one just south of the Rose Garden in Stanley Park, one on the north side of The Crescent, and one in the Mediterranean Section at VanDusen Botanical Garden.

'Leopoldii' – The variegated form with irregular yellow to cream splotches and streaks on the leaves. There is a very large tree along the west side of the pedestrian walkway beside the causeway in Stanley Park and several specimens at the Centennial Museum, along 15th Ave from Blanca St to Sasamat St, along 16th Ave from Mackenzie St to Trafalgar St, on the north side of 32nd Ave east of Oak St (at the Canadian Red Cross Society), and in the Maple Collection at VanDusen Botanical Garden.

Acer rubrum L.
Red Maple

Few trees can match the brilliant autumn colour of this common member of the eastern North American deciduous forest. A native to areas from Quebec to Florida, it is usually a good indicator of moist soils or even swampy places. It can grow to 40 m tall, but is quite variable in size and growth habit and is usually much smaller in cultivation. The leaves are dark green

above and glaucous grey beneath. They, too, are variable in size, lobing, and marginal teeth, but they generally have three or five predominant lobes. The tree might be confused with Silver Maple, but the leaves of Red Maple are less deeply lobed, less distinctly toothed, and not as silvery beneath. Small red flowers are produced in abundance in late winter or early spring before the leaves begin to emerge. In recent years it has become very popular as a street tree in many places, and there has been a confusing array of selected forms with narrow columnar or cone-shaped growth habit and usually very good red autumn colours.

The many typical wild forms in the city include specimens along the south side of Grandview Hwy between Rupert St and Skeena St, a row of relatively large ones (obviously seedlings, as they are quite variable) alternating with Purple-Leaved Plums along Brooks St between 47th Ave and 49th Ave, a street planting of large specimens of both Red Maple and Silver Maple along 39th Ave between Balaclava St and Carnarvon St (where the two similar species may be compared), and three relatively large wild forms in Queen Elizabeth Park west of the Rose Garden. •22

'**Columnare**' – The cultivar name used for the old, narrow forms of Red Maple. More recently, a number of narrow cultivars have been named and are popular as street trees. The most commonly planted one in our area is probably 'Armstrong.' There are long street plantings along Broadway, scattered between Main St and Clark Dr (especially between Clark Dr and Commercial Dr); along many blocks of Kingsway between Knight St and Baldwin St; and on Harwood St between Bute St and Nicola St.

'**Red Sunset**' – A relatively broad cultivar, with fairly good and uniform burgundy red colour in the autumn. It has been planted in the city in the past 15 years. There are long street plantings around the Vancouver Court House complex along Hornby St and Howe St from Robson St to Nelson St, and on the north side of 16th Ave between Wesbrook Mall and East Mall at UBC.

Cultivar Unknown – There is a very distinctive open-growth form, with sinuous upturned main limbs and few side branches. The leaves turn colour early in the autumn and are usually yellow, rather than the typical red of most red maples. It is an unknown cultivar or possibly an unnamed form. There are plantings along the south side of Cornwall Ave at Cypress St

and on the east side of Burrard St between 15th Ave and 16th
Ave, and a row on the west side of Carnarvon Park on the south
side of 16th Ave at Carnarvon St.

Acer rufinerve Siebold & Zucc.
Red-Veined Maple

This is another of the snake-barked maples. Its leaves usually
have pale edges which allow light to shine through when
viewed from below. Leaves are 3–5-lobed, but more often have 3
prominent lobes. The tufts of rusty hairs in the axils of the
veins on the lower side of the leaves best distinguish this species
from the other snake-barked maples. Long drooping chains
of green flowers are produced in spring, followed by chains of
small keys. There are usually some rusty hairs on both the
flowers and keys.

There are specimens on the northeast and west ends of the
Stanley Park Pitch & Putt Golf Course, and several in the Asian
Garden at UBC Botanical Garden and in the Sino-Himalayan
Garden at VanDusen Botanical Garden.

'Albolimbatum' – Although most individuals have white
margins on the leaves, this is the most extreme form with more
pronounced white margins and irregular light spots or flecks
of white and grey-green on the leaves. There is a large multi-
trunked individual on the east side of the hill in Queen
Elizabeth Park (just below the Seasons in the Park Restaurant).

Acer saccharinum L.
Silver Maple

This common eastern North American maple has rather large
deeply lobed leaves which are dark green on the upper surface
and silvery on the lower, hence its common name. Small
flowers in early spring (usually in February in our area) give the
trees a definite reddish cast before the leaves emerge. Autumn
leaf colour is variable from pale yellow to oranges and reds, but
not usually as showy as other maples. The trees eventually
become very large, growing to about 30 m tall, with deeply
grooved pale grey bark.

It is not very common here. There is a very long row through
the University Endowment Lands along University Blvd from
Blanca St to Wesbrook Cres, a street planting along Sasamat St
from 11th Ave to 12th Ave, two large specimens in the lawn area

NE of the Zoo near Lumberman's Arch in Stanley Park, and large ones at the Aberthau Cultural Centre at 2nd Ave and Trimble St and on the eastern side of Queen Elizabeth Park.

'Laciniata' – A cultivar with variably dissected and cut leaves. The only ones seen in the city are in Gastown along Water St between Carrall St and Cordova St, a row (mixed with some wild forms) along Nelson St from Stanley Park to Gilford St, and two trees on the north side of University Blvd at NW Marine Dr on the UBC campus.

Acer saccharum Marsh.
Sugar Maple

This is the common maple of eastern North America, growing naturally from Quebec to Florida and Texas. It becomes a large tree to 40 m tall and is the tree from which maple syrup and maple sugar are obtained commercially. It does not seem to be very happy in our climate, possibly because of our relatively cool, dry summers, or the wet, mild winters, or the acidity of our soil. Most of our trees are not as majestic as those of the East. It is easily mistaken for Norway Maple, which is one of our most common trees, but Sugar Maple is generally a taller, slender tree, with narrower leaves and fewer, longer lobes. Sugar Maple lacks the milky sap of Norway Maple and the petioles are generally shorter than the blades. On old trees the bark is more irregular, with large vertical plates, compared with the more uniform, narrow vertical plates of Norway Maple. Sugar Maple is probably the most spectacular tree for autumn colour in the East and it often gives a good show locally, if we have a mild, sunny autumn.

Probably the best specimens in Vancouver are in a row on the north side of Angus Dr from The Crescent to Granville St, and in The Crescent (growing with the similar Norway Maple, for comparison). There is a street planting, in various states of health, along Chancellor Blvd between NW Marine Dr and Acadia St in Point Grey and four young, but vigorous, trees behind the Chemistry Bldg on the UBC campus.

Acer shirasawanum Koidz. 'Aureum'
Golden Full-Moon Maple

The wild form of this Japanese maple is not known in cultivation in Vancouver except for a small specimen in the Asian Garden at UBC Botanical Garden. This golden-leaved form is very similar to *Acer japonicum,* and has long been listed in most references as a cultivar of that species. The minor characteristics that differentiate this species from *Acer japonicum* are that its petioles are not pubescent and the flowers are not pendulous. Otherwise, the two are very similar and difficult to distinguish. The cultivar 'Aureum' has very attractive pale yellow-green leaves.

There are two relatively large individuals north of the Rose Garden in Stanley Park and two small ones in the Asian Garden of UBC Botanical Garden.

Acer tataricum L.
Tatarian Maple

An attractive small tree native to southeastern Europe and western Asia, Tatarian Maple grows to about 9 m tall. The rounded to slightly lobed or sometimes definitely 3-lobed leaves are irregularly toothed and atypical of most familiar maples. They are thinner and less shiny than those of the similar Amur Maple (*Acer ginnala*). The creamy white flowers in rounded heads in May are attractive, as are the large red-pink keys produced in abundance in summer and autumn. The wings of the keys almost touch at their tips. It is not often seen in North American gardens.

The only ones found in the city include a beautiful, relatively large, round-topped tree on the UBC campus in a lawn just south of the offices of Resource Ecology (south of the Biosciences Bldg), and a long row of trees along Broadway between Renfrew St and Lillooet St.

Anacardiaceae – Cashew Family

Cotinus coggygria Scop.
Smoke Tree

This central and southern European large shrub or small tree reaches 4 m tall and is seen in gardens as an ornamental, usually in its purple-leaved form. The common name comes from the smoky effect of large puffs of yellow-green flowers and, especially, the clouds of plumy fruits. The trees are attractive from the time the flowers are out in late May or early June, throughout the summer, and into the winter. Long-petioled leaves are oblong or spoon-shaped, often turning orange or red in the autumn.

The typical wild form with dull green leaves is rarely cultivated. There is a small single-trunked tree along the east side of Blanca St between 2nd Ave and 3rd Ave, two large tree forms on the south side of 11th Ave between Blanca St and Tolmie St, and a large multi-trunked tree on the west side of Angus Dr between Laburnum St and 64th Ave. •23

'Purpureus,' Purple Smoke Tree – This is the common form seen in gardens. It has very rich purple-black leaves that turn a lighter red-purple in the autumn. There are many around the city, but few truly tree forms. The larger ones include those on the SE corner of Laburnum St and 62nd Ave, on the NW corner of Collingwood St and 29th Ave, and on the north side of 12th Ave just west of Woodland St; and there is a grove in the courtyard of Buchanan Bldg at UBC. There are both green and purple forms in Queen Elizabeth Park along Cambie St just north of 33rd Ave.

Cotinus obovatus Raf.
American Smoke Tree

Much larger in all aspects, except for the flower clusters, and much rarer in cultivation than *Cotinus coggygria*, this attractive tree grows to about 10 m tall. It is native to the southeastern United States from Tennessee and Alabama to Missouri and Texas, but is much hardier than the natural range would suggest. The rounded to obovate leaves are about twice as large as those of the European species. The leaves are bronze when they first emerge in spring and often turn brilliant orange to red in autumn. Narrow panicles of small green flowers are borne in late spring. It is a very picturesque tree but not often seen in cultivation in the West.

The only tree seen in the city is a very beautiful, relatively large one in a yard on the [illegible] corner of [illegible] Ave and Highbury St

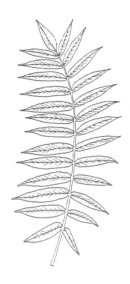

Rhus typhina L.
Staghorn Sumac

Although closely related to poison ivy, this attractive small tree from the eastern North American deciduous forests is not at all poisonous to the touch and is a pleasant addition to the garden throughout the year. The branch structure, with thick, brown furry twigs, gives the tree its common name.

Large pinnately-compound leaves with marginal teeth and a glossy upper surface are attractively dark green in summer, but especially good in the autumn when they turn brilliant oranges and reds. It is among the most colourful of our locally grown trees in autumn, especially following a dry summer. It is usually fully coloured by mid- to late September, earlier than many other trees. There are separate male and female trees, with the males producing slightly larger, more open panicles of tiny yellow-green flowers in June to early July. These produce pollen and then drop. The flowers on female trees form the tight, elongated, fuzzy, dull red seed heads that are noticeable on the ends of branches from mid summer well into the winter. The small trees are very common in Vancouver, but may go unnoticed until the autumn colour begins.

There is a very large, old, multi-trunked tree in a garden on the south side of King Edward Ave just west of Ontario St, one on the sw corner of Willow St and King Edward Ave, one on the NE corner of Stephens St and 15th Ave, several on the south side of 12th Ave just east of Arbutus St, and several in the vicinity of 23rd Ave and 24th Ave, and Ash St and Heather St.

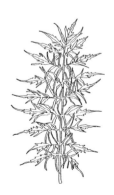

'Laciniata,' Cut-Leaved Staghorn Sumac – This is a smaller tree than the typical wild form and often remains shrubby. The branches are more contorted, and the leaves are dissected and fern-like. It is not nearly as common locally as the wild form, although it is more attractive. There are many small ones around the city, but among the larger tree forms are those on the sw corner of 23rd Ave and Dunbar St, a nice pair on the east side of Acadia Rd just north of College Highroad, one on the east corner of Gilford St and Comox St, groves of trees in the Dry Section and in the Heather Garden at VanDusen Botanical Garden, and a nice picturesque specimen on Memorial Rd north of the Old Administration Bldg at UBC. •24

Aquifoliaceae – Holly Family

Ilex – Hollies

We generally think of the hollies as being spiny evergreen shrubs or trees with red berries used at Christmas, but there are many other forms, mostly shrubs, that are either deciduous or evergreen, often with black berries. The plants are usually either male or female, with small, white, 4-petalled flowers. The only tree-form holly cultivated locally is the English Holly and its hybrids. The common eastern North American tree-form holly is American Holly (*Ilex opaca*), with much duller leaves than the English Holly. It needs longer, hotter summers than we can provide, so is very rarely cultivated in our area. There are a number of other Asian species cultivated at UBC Botanical Garden and many cultivars of English Holly and other species in the Ilexetum (Holly Arboretum) at VanDusen Botanical Garden, an official test site for the American Holly Society.

Ilex leaves
1 *I. aquifolium*
2 *I. aquifolium*
3 *I. aquifolium*
   variegated form
4 *I. ×altaclerensis*
5 *I. ×altaclerensis*

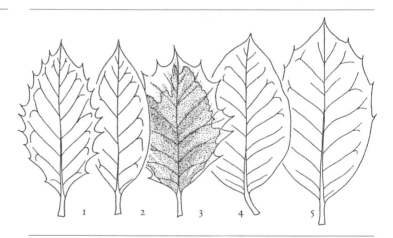

Ilex aquifolium L.
English Holly

This broad-leaved evergreen is very commonly cultivated in our area but is usually clipped and kept shrubby, rather than allowing it to grow into a tree. It is a native of Europe, North Africa, and western Asia, where it grows to at least 15 m tall. The smooth grey bark and green branches with very waxy, dark green, usually spiny, leaves makes it one of our most distinctive shrubs. It is one of the plants used traditionally at Christmas, especially the red-berried branches from female trees. The plants are either male or female. Older trees, especially female,

tend to become less prickly as they grow older. The bright red berries are attractive to birds that carry them away, resulting in seedlings often being seen in open forests throughout the Lower Mainland. English Holly is a commonly naturalized part of the flora of Pacific Spirit Regional Park and the University Endowment Lands, for example. It is one of those plants that shows a great deal of seedling variation in leaf size, shape, spines, and colours. Also, an individual branch may mutate and become different from the remainder of the tree. Cuttings from the branch may be rooted to produce a new plant that will usually retain its different character. This has resulted in the development of many different cultivated forms of English Holly, for example, forms with yellow or orange fruits, of many different sizes, shapes, and variegations of leaves, and of varying degrees of spininess.

Typical wild English Holly trees of the large tree form around the city include several large ones (mostly females) in a street planting on the SE corner of 6th Ave and Blenheim St, several large ones in The Crescent, two large females on the south side of 41st Ave just west of Blenheim St, a nice male on the NE side of Narvaez Dr at Puget Dr, two females on the SW corner of Collingwood St at 21st Ave, and a heavy-fruiting individual on the north side of 41st Ave west of Blenheim St.

Variegated forms – There are many differently named cultivars with variegated leaves. There are two large variegated (white-edged leaves) females on the south side of Tatlow Park along 3rd Ave just west of Macdonald St and one on the NE side of Burnaby St between Cardero St and Bidwell St. Yellow-edged female trees can be seen on the east side of the Stanley Park Dining Pavilion, on the south side of 11th Ave between Tolmie St and Blanca St, and on the north side of 5th Ave between Balsam St and Larch St; and there are several yellow-edged female trees in a garden on the SW corner of Chancellor Blvd and Acadia Rd (with a row of typical green forms).

Ilex × altaclerensis (Hort. ex Loud.) Dallim.
Highclere Holly

This tree is a hybrid between *Ilex aquifolium* and *Ilex perado,* the first parent being very common here, while the second is from the Canary Islands and is not hardy this far north. The hybrid is becoming more popular in recent years and there are many small ones in the city. It is said to have originated at Highclere, the estate of the Earl of Carnarvon near Newbury, Berks, England. It is variable, but similar in appearance to

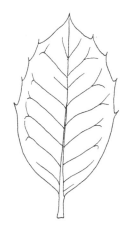

English Holly except that the leaves are generally larger, shinier, and usually spineless or nearly so; the stems are thicker and often quite purple; and the berries are usually larger. It may be very similar to, and difficult to distinguish from, mature forms of the English Holly.

There are a number of small cultivars in the Holly Collection at VanDusen Botanical Garden, a large female on the west side of Connaught Dr between 35th Ave and 36th Ave (probably the cultivar 'Camelliifolia'), a male on the NE side of Narvaez Dr at Puget Dr, and a number of relatively large ones at UBC, including a female on the north side of University Blvd behind the Education Bldg and two females and a male on the west side of the Old Armoury along West Mall.

Araliaceae – Aralia Family

Aralia elata (Miq.) Seem.
Japanese Aralia or Japanese Angelica Tree

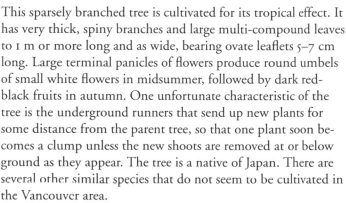

This sparsely branched tree is cultivated for its tropical effect. It has very thick, spiny branches and large multi-compound leaves to 1 m or more long and as wide, bearing ovate leaflets 5–7 cm long. Large terminal panicles of flowers produce round umbels of small white flowers in midsummer, followed by dark red-black fruits in autumn. One unfortunate characteristic of the tree is the underground runners that send up new plants for some distance from the parent tree, so that one plant soon becomes a clump unless the new shoots are removed at or below ground as they appear. The tree is a native of Japan. There are several other similar species that do not seem to be cultivated in the Vancouver area.

There are a number of small, shrubby specimens and a few large tree-like forms around the city, including a group by the causeway, at the south entrance into Stanley Park; three specimens as street trees on the south side of 48th Ave west of Oak St; and single-trunked specimens on the south side of 31st Ave between Blenheim St and Balaclava St, on the NE corner of 18th Ave and MacKenzie St, two on the north side of 26th Ave between St. George St and Balkan St, on the south side of 7th Ave between Trutch St and Blenheim St, and in the courtyard of Buchanan Bldg between Main Mall and East Mall on the UBC campus. •25

Arecaceae (Palmae) – Palm Family

Trachycarpus fortunei Wendl.
Windmill Palm

There is little doubt that this is the hardiest member of this very important family of tropical trees. The Windmill Palm is a native of Japan, China, and the Himalayas, where it is usually exposed to frost and snow in winter. The broad, dark green, fan-like leaves are borne atop a thick trunk, as tall as 12 m in nature and covered with dark brown hair-like fibres, resembling a frayed piece of dark burlap. It is a relatively fast-growing tree. Large panicles of yellow flowers are borne on short stalks among the leaves in summer, followed by blue-black round fruits. Local specimens have been known to produce viable seeds.

A number of specimens have survived in the area with little or no protection, although many others were killed in recent severe winters. There seems to be much variation in hardiness between individuals, probably not only in relation to where they are planted but also due to the origin of the seed and, probably, genetic differences. Admittedly, this palm does best in our area when given a warm, sunny, protected location.

The largest specimens in the city that are readily seen include a very tall one (that is probably the oldest one in the city) among the trees south of the monkey cages at the Zoo in Stanley Park, one on Point Grey Rd at Bayswater St, several at VanDusen Botanical Garden (at the entrance and in the Meditation Garden), several (including a relatively large specimen) in a garden on the SE corner of 40th Ave and Manitoba St, a very nice specimen in a back garden (visible from the lane) between 59th Ave and 60th Ave just west of Granville St, and nice young specimens against houses on the NE corner of 1st Ave and Stephens St, the NW corner of 10th Ave and Glen Dr, and on the NW corner of 56th Ave and Vivian Dr.

Betulaceae – Birch Family

Alnus – Alders

The alders are not usually valued much as ornamentals. Some, especially our native Red Alder, are generally thought of as weed trees, because they are among the first trees to recolonize bare ground. The simple, oval to egg-shaped leaves are usually double-toothed or sometimes very slightly lobed. Long rounded winter buds open in early spring, producing long pendulous male catkins, whose pollen is one of the greatest local causes of hay fever. Female catkins are very short and, after pollination, grow into oval, brown cone-like structures containing tiny winged fruits. Autumn colour is usually very dull.

Alnus leaves
1 *A. crispa*
2 *A. glutinosa*
3 *A. glutinosa* 'Imperialis'
4 *A. incana*
5 *A. rubra*
6 *A. tenuifolia*

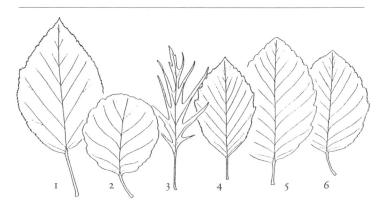

Alnus crispa (Ait.) Pursh ssp. *sinuata* (Regel) Hult. (*Alnus sinuata* Regel)
Sitka Alder

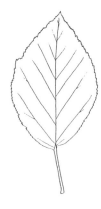

A native of Alaska to northern California and Montana, this alder has a distinctive branching habit, with several trunks arching out from the base to form broad bowl-shaped clumps. It is often small and shrubby in nature. The leaves are sharply double-toothed and the edges are flat, not turned under as they are on Red Alder. The catkins are very long and prominent in early spring. It is not generally considered of any ornamental value, but is sometimes grown in botanical collections.

There is a large, very typical tree with many trunks in the Old Arboretum at UBC. It is larger than is usually seen in the wild because it is growing out in the open with no competition. There are smaller specimens in the BC Native Garden of UBC Botanical Garden.

Alnus glutinosa (L.) Gaertn.
Black Alder or European Alder

This Eurasian tree has very broad, almost round leaves and a
sticky texture to the buds and leaves, especially when they first
emerge in spring, thus the species name *glutinosa,* meaning
sticky. It is easily distinguishable from our other locally grown
alders as it is the only one with leaves that are broadest near the
tip rather than near the middle or the base of the leaf.

The only specimen of the wild form of this tree seen in the city
is one on the NW corner of Queen Elizabeth Park at Cambie St
and 29th Ave.

'Imperialis' – This strange-looking cultivar looks almost
nothing like the wild form. It has leaves with little more than
the major veins and a bit of leaf tissue between them. There are
two trees on the NW corner of Larch St and Cornwall Ave, a
specimen across the road from the south side of the Pitch &
Putt Golf Course in Stanley Park, and two trees at the
beginning of the Rhododendron Walkway at VanDusen
Botanical Garden.

Alnus incana (L.) Moench.
White Alder

A Eurasian tree growing to 20 m tall, this alder closely resembles our abundant native Red Alder, except that it has glaucous
or grey undersides to the leaves; the leaf edges are flat, not
rolled under; and the catkins are paler, usually yellow, before
opening and shedding pollen. The leaves are also narrower with
more distinct, sharp points. Catkins tend to open earlier in the
spring than those of Red Alder. It may be distinguished
from the rarely cultivated, but similar, Mountain Alder by the
narrower leaves and distinctly narrower, more sharply pointed
teeth of the White Alder.

It is very rare here. The only trees seen are one at UBC between
International House and Panhellenic House just off NW Marine
Dr, one at West Mall (growing with a similar *Alnus tenuifolia*),
and a small one in the SE corner of Queen Elizabeth Park that
has pale leaves and appears very unhealthy. This one is probably
the cultivar 'Aurea.'

Alnus rubra Bong.
Red Alder

This is by far our most common native deciduous tree, growing from southern Alaska to California. It is usually considered a weed tree because it is so common, reseeds so prolifically, and grows so rapidly. It is one of the first plants to colonize any bare soil, forming dense thickets of seedlings and, later, saplings. Trees have thin, pale grey bark. The name Red Alder comes from the reddish colour of its freshly cut wood and from the dull red male catkins. Catkins are borne in abundance on the bare branches in late winter and early spring, giving the trees a definite red cast that is visible at long distances. Male catkins generally elongate to about 5–10 cm and become dull yellow as they shed masses of pollen, usually in March. The pollen is the cause of one of the most common spring allergies in this area. Smaller female catkins are borne in clusters of 3–5. They become hard, woody, cone-like structures, superficially resembling a small conifer cone. These hang on the trees throughout most of the winter. The large, elongated winter leaf buds are also distinctive in winter. Dull green leaves are distinctly pinnately veined, have large rounded teeth, and the edges are slightly rolled under. This characteristic is a good way to separate Red Alder from other alder species. The leaves become duller green in the autumn and drop late in the season without any show of colour.

It is usually not cultivated here, because it is so common and considered weedy, but may be seen in abundance in any native forest, such as Pacific Spirit Regional Park, the University Endowment Lands, Stanley Park, or Queen Elizabeth Park. There are two large specimens by the pedestrian underpass on the NW corner of Stanley Park Pitch & Putt Golf Course, a number of very large trees by the concession stand at Third Beach in Stanley Park (including a 4-trunked specimen that is recorded as the largest Red Alder in Canada), and a long row of large specimens on the south side of 33rd Ave between Oak St and Willow St.

Alnus tenuifolia Nutt.
(*Alnus incana* [L.] Moench ssp. *tenuifolia* [Nutt.] Breit.)
Mountain Alder

This is the common alder in British Columbia east of the Coast/Cascade Mountains. Its broad leaves are very similar to those of the Red Alder, except that the edges are flat, not rolled

under, and there are soft hairs on the petioles and along the veins beneath. It is not particularly ornamental and is rarely cultivated.

The only trees known in the city include a large one at UBC between International House and Panhellenic House just off NW Marine Dr, one at West Mall (growing with a similar *Alnus incana*), and a few smaller ones in the BC Native Garden of UBC Botanical Garden.

Betula – Birches

A majority of cultivated birches are raised for their smooth white or pale bark, which peels off in large papery sheets. Usually rather small trees, birches are among the dominant vegetation at the northern extremes of deciduous trees. The leaves are relatively small and oval, rounded or almost diamond-shaped, usually very thin, and distinctly toothed. Short erect female catkins and long drooping male catkins are produced in spring. The female catkins fall apart at maturity, unlike those of the related alders whose female catkins become hard and woody and remain intact at maturity.

Betula leaves

1 *B. albo-sinensis*
2 *B. papyrifera*
3 *B. populifolia*
4 *B. pendula*
5 *B. pendula* 'Youngii'
6 *B. pendula* 'Dalecarlica'
7 *B. pubescens*
8 *B. utilis*

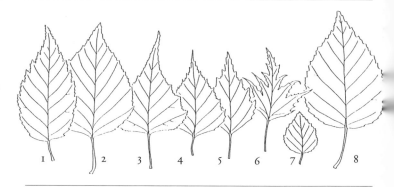

Betula albo-sinensis Burk.
Chinese White Birch

Without a doubt the Chinese White Birch is one of the nicest of deciduous trees and certainly deserves wider recognition. Although variable, the beautiful, pale coppery pink bark, peeling away in great sheets, is the distinguishing feature of this Chinese birch. The leaves are paler green than the similar Himalayan White and other birches. There is a line of pubescence along the midrib beneath. It does become a very large tree (to 35 m tall) in its native habitats in China, but all of the specimens in the city are relatively young and, as yet, small trees.

There are five beautiful specimens on the sw corner of the
Stanley Park Pitch & Putt Golf Course, two at the entrance to
the rest rooms for Queen Elizabeth Park Pitch & Putt Golf
Course, one by the Electrical and Mechanical Engineering Bldg
on East Mall opposite the parkade at UBC, several in the Asian
Garden of UBC Botanical Garden, and several in the Sino-
Himalayan Garden at VanDusen Botanical Garden.

Betula papyrifera Marsh.
Paper Birch, White Birch, or Canoe Birch

This native tree has a broad oval outline and papery, white
bark, making it an attractive ornamental, but it is not often
cultivated, probably just because it is native. It is sometimes
interplanted with European White Birch (*Betula pendula*), from
which it may be distinguished by its larger leaves and less
drooping habit. The leaves are broadly oval with irregular teeth
around the margins. There are often distinct bumps along the
slender twigs that can be felt if the twigs are gently pulled
through the fingers. Slender, wiry male catkins are noticeable
from summer until they elongate in spring. Female catkins are
larger in diameter and fall apart upon ripening.

There is a magnificent mature specimen in the Old Arboretum
at UBC that has been out on its own, with little competition,
and has developed a very broad shape, very unlike the typical
trees seen in the wild. There are also several between Cecil
Green House and the Museum of Anthropology at UBC, street
trees on the north side of Lagoon Dr between Robson St and
Haro St, a nice row alternating with European Copper Beech
(*Fagus sylvatica*) along 13th Ave between Dunbar St and
Waterloo St, three large ones on the east side of Willow St
between 14th Ave and 15th Ave, and a grove of lovely trees on
the east side of Queen Elizabeth Park. •26

Betula pendula Roth (*Betula alba* L.)
European White Birch or Weeping White Birch

This attractive deciduous tree is cultivated for its white bark,
usually graceful habit, and pendulous branches, making it one
of the most popular street or park trees. It is native to Europe
and Asia Minor, reaching 20 m in the wild. The white bark
brightens our winter days, and the pendulous habit is attractive
throughout the year. There is a great deal of variation in habit,
varying from weeping to nearly erect branches. Leaf colour is

often a good gold in the very late autumn and it is one of the last trees to colour in our area. The leaves are small enough not to cause a raking problem when they drop.

It is very common here as a street tree. Among the street plantings are rows on both sides of 27th Ave from Dunbar St to Balaclava St (these trees have varying degrees of pendulous branches), along 35th Ave from Dunbar St to Crown St, along the south side of 33rd Ave from Granville St to Oak St, on 46th Ave from Main St to Ontario St, and along East Blvd from 49th Ave to 52nd Ave.

'Dalecarlica,' Cut-Leaf Weeping Birch – A commonly grown cultivar with distinctive, dissected leaves. Unfortunately, it does not do well here and often has a short, dense, twiggy growth so that it usually does not look as good as the typical wild form or other cultivars. There is a relatively good group west of the Rose Garden in Queen Elizabeth Park, and street plantings on Dunbar St between 12th Ave and 13th Ave, along Angus Dr between 37th Ave and 41st Ave, and along 5th Ave from Sasamat St to Trimble St.

'Fastigiata' – This cultivar has upright sinuous branches forming an irregular elliptic shape. There are street plantings on 49th Ave from Tyne St to Nanaimo St and on Harwood St between Thurlow St and Bute St; there are several large trees west of the Quarry Garden in Queen Elizabeth Park and three east of the main entrance to Vancouver General Hospital on 12th Ave at Heather St.

'Purpurea,' Purple-Leaved Weeping Birch – A rarely grown cultivar, at least in Vancouver, with dark red-purple leaves that contrast well with the white trunks. There is a group of trees on the south side of 33rd Ave just east of Cambie St at the entrance to Queen Elizabeth Park, two in the Heather Garden at VanDusen Botanical Garden, and one against the south wall of Lasserre Bldg on the NW corner of Main Mall and Memorial Rd at UBC.

'Youngii,' Young's Weeping Birch – Sometimes called the 'Mop-top Birch' by those who do not like its form. It is usually grafted on a standard trunk about 2 m high and the branches then weep strongly, forming an umbrella shape. There are two on the NE corner of 14th Ave and Birch St, two relatively large ones on the west side of Peveril Ave near Manitoba St, another large one on the SE corner of Dogwood Ave and Cartier St, a row on the north side of 3rd Ave between Balsam St and Larch St, a group of four in front of the Parks Board Office on Beach Ave, several in front of the Main Library at UBC, and one on the east side of Waterloo St between 13th Ave and 14th Ave.

Betula populifolia Marsh.
Grey Birch

This northeastern North American birch is very similar to the White Birch (*Betula pendula*), but the leaves of Grey Birch are usually a bit larger and have much more elongated tips. The bark is usually not quite as white as that of its European relative, and the twigs are usually noticeably glandular-warty, feeling rough to the touch.

This seems to be a rarely cultivated birch in Vancouver, but it is often difficult to distinguish from other similar birches and may be more common than is thought. There are several around Sawyers Lane and Greenchain in False Creek.

Betula pubescens J.F. Ehrh.
Downy Birch or Silver Birch

This European native is similar to, and often confused with, the much more commonly cultivated European White Birch (*Betula pendula*). Downy Birch has grey to white bark and less pendulous branches, with pubescent young twigs. The dark green leaves (to 5 cm long) have a neat teardrop shape and are also pubescent below. The trees grow to about 20 m tall.

Locally, Downy Birch seems to be uncommon, but it is easily confused with the very common European White Birch so there may be more of them than it seems. There is a nice specimen on the NW corner of Buchanan Bldg at UBC, two on the north side of Sawcut at Millyard in False Creek, two on the west side of the tennis courts plus others in a grove north of 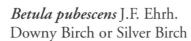 McNicoll Ave in Kitsilano Beach Park, and at least one on the SE corner of the Stanley Park Pitch & Putt Golf Course.

Betula utilis D. Don var. *jacquemontii* (Spach) Winkler
(*Betula jacquemontii* Spach)
Himalayan White Birch or Kashmir Birch

This is by far the most attractive of all white-barked birches and deserves to be cultivated much more that it is. The bark is the whitest of all birches and this, combined with very dark green leaves and a broad teardrop shape, makes it one of the best of small deciduous trees. A native of the Himalayas, it ultimately reaches about 18 m tall. Ovate, dark green leaves are about 5–7 cm long, with 7–9 pairs of veins that are hairy beneath. The

variety with the purest white bark has sometimes been
considered a separate species, *Betula jacquemontii*, but recent
studies on the group suggest that it is better regarded as a
variety of the more widespread Himalayan species, *Betula utilis*.

There are a number of nice individuals around the city in parks
and public gardens, but very few otherwise. There is an individual on the NE corner of 20th Ave and Collingwood St; five
on the NE corner of Oak St and 7th Ave; and several around the
Pitch & Putt Golf Course in Stanley Park, in the Asian Garden
at UBC Botanical Garden, and in the Sino-Himalayan Garden
at VanDusen Botanical Garden. The best planting is a grove
to the west of the Heather Garden at VanDusen Botanical
Garden. •27

Carpinus betulus L.
European Hornbeam

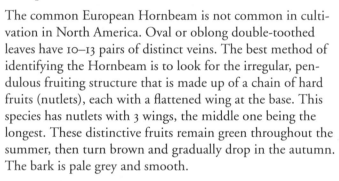

The common European Hornbeam is not common in cultivation in North America. Oval or oblong double-toothed
leaves have 10–13 pairs of distinct veins. The best method of
identifying the Hornbeam is to look for the irregular, pendulous fruiting structure that is made up of a chain of hard
fruits (nutlets), each with a flattened wing at the base. This
species has nutlets with 3 wings, the middle one being the
longest. These distinctive fruits remain green throughout the
summer, then turn brown and gradually drop in the autumn.
The bark is pale grey and smooth.

The wild form is not nearly as common as the erect cultivar
'Fastigiata.' The few seen in the city include a street planting on
the east side of Prince Edward St from King Edward Ave to
28th Ave, two large ones on the south side of Stanley Park near
the turnoff for the Zoo parking lot, a group of seven trees on
the north side of Queen Elizabeth Park, and an old one at UBC
between the garden area NW of the Main Library and Main Mall.

'Fastigiata' – The common form seen here, with very
distinctive erect branches, is particularly noticeable in winter.
There are street plantings along 1st Ave from Clark Dr to
Victoria Dr, on the NE corner of Oak St and 49th Ave, along
Beach Ave from Thurlow St to Jervis St, on the north side of
6th Ave from Alder St to Willow St, and along Trafalgar St
from 4th Ave to Broadway.

'Incisa' – An ugly form with variably cut and dissected leaves, making it look as if it has been chewed upon by some insect. There are often limbs with both dissected and normal, broad leaves on the same twigs. The only ones seen in the city are three trees on the west side of Carnarvon Park south of 16th Ave at Carnarvon St.

Carpinus caroliniana Walt.
American Hornbeam

Very similar to the European Hornbeam, this American relative differs in some small characteristics of the leaves and fruits. The leaves are generally smaller, and the fruiting clusters and the wings of individual fruits are not as long as those of the European Hornbeam. The tree is usually not as broad as the wild forms of European Hornbeam. It is native throughout much of the eastern deciduous forest, sometimes reaching 15 m tall. The smooth bark on sinuous limbs gives it a common name of Muscle-Wood.

The only ones known in the city are several slender trees planted in 1985 at the Expo '86 site, along the seawall at the foot of Davie St.

Carpinus tschonoskii Maxim.
Tschonoski's Hornbeam

This small deciduous tree grows to 15 m tall and is native to northern Asia. It may be distinguished from the other locally cultivated hornbeams by its densely pubescent young twigs and by the bracts surrounding the nutlets, which have a single lobe on each wing rather than the 3-lobed wings of the other species. The smooth grey bark and sinuous branches are distinctive in winter.

It is very rarely cultivated. There are three old multi-trunked trees on the south side of the Graduate Student Centre along Crescent Rd at UBC.

Corylus – The Hazelnuts or Filberts

Most of the ten or so species of *Corylus* are shrubs or small trees cultivated for their sweet nuts. A few, such as the Turkish Hazel (*Corylus colurna*), may become very large trees, but these are seen locally only in botanical collections, such as UBC Botanical Garden and VanDusen Botanical Garden. The native Beaked Hazelnut (*Corylus cornuta*) is a fairly common native shrub or small tree in the forests of Stanley Park, Pacific Spirit Regional Park, and the University Endowment Lands. It is not known to be cultivated in Vancouver. Most *Corylus* species are difficult to distinguish unless the nuts are visible, as the shape and texture of the bracts surrounding the nuts are distinctive for each species. Hazelnuts are most noticeable in winter or early spring when they are in flower because of their long, pendulous, yellow male catkins.

Corylus avellana L.
Common Hazelnut or Common Filbert

This European shrub or small multi-trunked tree grows to about 5 m tall and is raised commercially for its edible nuts. It is one of the earliest of trees to flower, with the long, pendulous, yellow male catkins often elongating and shedding pollen in December or January in Vancouver. The female flowers are very small, merely a red fringe that barely extends beyond the buds. The leaves are rounded with toothed edges, and they are a bit smaller than those of the other hazelnuts cultivated here. The dissected, fringed bracts surrounding the nuts are shorter than the nuts.

It is a commonly cultivated nut tree in gardens. There are many, mostly small, shrubby specimens in the city. Larger multi-trunked tree forms include specimens on the NE corner of 11th Ave and Spruce St, east of the Stanley Park Dining Pavilion, and west of MacMillan Bldg on the UBC campus; there are three large trees on the SE corner of 1st Ave and Highbury St and four along Trimble St, between 7th Ave and 8th Ave on the west side of West Point Grey Park.

'Contorta,' Harry Lauder's Walking-Stick or Contorted Hazel – A curious small tree with very twisted twigs, catkins, and leaves. Its bare branches are attractive in winter and are popular with flower arrangers. However, it is not very pleasing in summer with its twisted leaves. There is a small tree on the south side of 5th Ave between Bayswater St and Balaclava St, one on the north side of 6th Ave between Templeton Dr and Garden Dr, one on the west side of Dumfries St between 55th Ave and 57th Ave, and large ones in the Alpine Garden of UBC Botanical Garden and in the Heather Garden of VanDusen Botanical Garden.

Corylus maxima Mill.
Great Hazelnut or Great Filbert

Generally larger in all respects than the Common Hazelnut, this species is somewhat less common locally. The leaves tend to be softer on the upper surface and are generally larger and broader than those of the Common Hazelnut, if the two are compared closely. The larger nuts are borne singly or sometimes in groups of two or more, with bracts flaring and extending beyond the end of the nut (in the green form) or surrounding the nuts and becoming a long beak, twice the length of the nuts and flaring at the tip (in the purple form). There are sharp, fine bristles at the base of the bracts.

The typical wild form with green leaves is rarely seen here. The purple-leaved cultivar is more common. There are green-leaved trees in the lawn area south of the Stanley Park Pitch & Putt Golf Course, one by the causeway on the west side of the pedestrian bridge overpass in Stanley Park, and three in the Food Garden at UBC Botanical Garden.

'Purpurea,' Purple Filbert – This cultivar has very dark, almost black-purple leaves and is planted more often than the green forms because of this attractive feature. The bracts surrounding the nuts and the long, pendulous male catkins are also purple. The plant serves a double function, as an ornamental large shrub or small tree and as a nut producer. It is fairly commonly planted around the city, usually as a large bushy shrub. There are multi-trunked trees on the north side of the Aberthau Cultural Centre at the NE corner of 1st Ave and Trimble St, on the north side of the Math Annex at Main Mall and Agriculture Rd at UBC, on the north side of 14th Ave between Balsam St and Larch St, and in Stanley Park on the west side of the Pitch & Putt Golf Course and in the perennial beds west of the Rose Garden.

Ostrya virginiana (Mill.) C. Koch
Hop Hornbeam or Ironwood

This smooth-barked deciduous tree, growing to about 20 m tall, is common in a wide area in the eastern deciduous forest from Nova Scotia and Minnesota south to Florida and Texas. When in leaf, it is difficult to distinguish from the related genus *Carpinus*. Oblong-ovate leaves are 10–12 cm long, and have short petioles, distinct pinnate veins, and toothed margins. The fruits are hop-like, as the common name suggests, in 5 cm long drooping clusters. Each nutlet is completely enclosed by a papery, bladder-like envelope, with long, sharp hairs at the base. The fruits are pale green at first, becoming soft brown before they break apart and drop in the autumn. Each nutlet of *Carpinus* has a bract that is more or less flat, not enveloping the nutlet.

The only specimens seen in the city are three rather large individuals in the lawn just outside the fence on the north side of the Queen Elizabeth Park Pitch & Putt Golf Course.

Bignoniaceae – Bignonia Family

Catalpa

Catalpas are among a very few hardy members of a widespread tropical family. Their large leaves and large, ruffled flowers are indeed tropical-looking. The pale green, heart-shaped leaves are attractive throughout the summer and as they become yellow in the autumn; but catalpas are among the latest of trees to leaf out in the spring. Most species have fragrant, white, trumpet-shaped flowers in midsummer. Close inspection shows that they are spotted with purple and lined with 3 yellow bands that change to orange-red as the flowers fade. The flowers are followed by long bean-like pods that hang on the trees throughout the winter. Not all trees flower every year, and there is a great deal of variation in the numbers of flowers from tree to tree.

Catalpa leaves and flowers

1 *C. bignonioides*
2 *C. ×hybrida*
3 *C. ovata*
4 *C. speciosa*

Catalpa bignonioides Walt.
Common Catalpa

This tree is very similar to the Western Catalpa (*Catalpa speciosa*), but does not become as large, usually growing to only about 20 m tall. It is more southern in origin, being found in the wild from Georgia to Florida and Mississippi, but it is hardy and widely cultivated, although not quite as common in Vancouver as Western Catalpa. The large, pale green, ovate leaves are less long-pointed than those of Western Catalpa. The leaves are reported to be foul-smelling when bruised, but this characteristic is not very reliable in our climate. Flowers are borne several weeks later than Western Catalpa (usually late July to August), in larger, longer clusters (usually much longer than wide), but the individual flowers are smaller (3–4 cm wide). The fruits are also not quite as long and are more slender (less than 1 cm wide).

There is a tree at the north end of the pedestrian overpass by the causeway in Stanley Park, and one on the west side of Queen Elizabeth Park, below the Rose Garden. Mixed plantings with Western Catalpa are found: in front of the Canadian National Railways Station between Station St and Main St, as street plantings along 9th Ave from Vine St to Trafalgar St, on 10th Ave from Stevens St to Blenheim St, along 12th Ave from Trafalgar St to Vine St, on Angus Dr from 33rd Ave to 37th Ave, and on Nelson St from Gilford St to Bidwell St. •28, 29

'Nana,' Umbrella Catalpa – A dwarf form that is usually grafted on a standard trunk. The trees have smaller leaves and a dense rounded form like a green lollipop. They are often pruned back to near the graft to produce an even tighter growth habit. It is widely cultivated in some places, but is rare here (fortunately, in the author's opinion!). There are two trees by the Rose Garden and one in the Children's Garden at VanDusen Botanical Garden, and two as a street planting on the north side of 39th Ave just west of Carnarvon St.

Catalpa × hybrida Hort. (*Catalpa × erubescens* Carr.)
(*Catalpa bignonioides* Walt. × *Catalpa ovata* G. Don)
Hybrid Catalpa

This hybrid is variable in leaf size and colour, but the young leaves are usually dark purple, becoming pale green as they mature. The flowers are in large, long clusters, much like those of the Common Catalpa, but they have more purple spots and streaks, giving them a much darker look at a distance. The trees

also tend to have a few leaves with 3-5 lobes, like the Chinese Catalpa (*Catalpa ovata*). The beans are slender, less than 1 cm wide, and 18–22 cm long.

It is rarely cultivated here. There are two trees in Queen Elizabeth Park on the north side (with one of the parents, Chinese Catalpa), a large tree on the south side of 12th Ave just east of Arbutus St (behind a building, but visible from 12th Ave), and several on the east side of Ash St between 13th Ave and 14th Ave (with both parents). •30

Catalpa ovata G. Don
Chinese Catalpa or Yellow Catalpa

This Chinese species is distinctive among our local catalpas, in that it has yellow-green flowers with purple spots. The colour is noticeable at a distance, compared with the other species with much whiter flowers. The leaves tend to be more often 3- or 5-lobed. Flower clusters are much longer than wide, and the inflorescences and individual flowers are the size of the Common Catalpa.

It is rare here. There are two trees on the north side of Queen Elizabeth Park (growing with two trees of the offspring, Hybrid Catalpa) and one on the corner of 14th Ave and Ash St (the southernmost tree in a row of mixed Hybrid Catalpa and Common Catalpa).

Catalpa speciosa Warder ex Engelm.
Western Catalpa

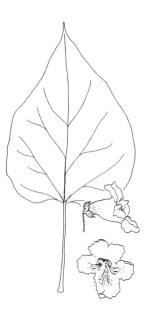

The most commonly cultivated catalpa in Vancouver is the Western Catalpa. It is native to the east-central United States and becomes a larger tree than the Common Catalpa, often reaching 30 m or more tall. The large, pale green, ovate leaves have long slender tips and are reported to have no smell when bruised. Individual trees are variable as to their time of flowering. The flowers are produced as early as mid-June, but the peak of flowering is in July and is finished in late July, just as the flowers of the Common Catalpa begin to open. Close inspection shows that the flowers have ruffled petals and beautiful spots and streaks of purple with three yellow to orange blotches. They are less spotted than those of the Common Catalpa and larger (5–6 cm wide), but there are fewer in each inflorescence and the inflorescences are usually as wide as they are long. The fruits are the largest of the catalpas, 25–30 cm long and more than 1 cm thick.

It is common here. Among the good specimens of this tree are those along the pedestrian walkway parallel to the causeway in Stanley Park, on the south side of Buchanan Bldg from the north end of Main Mall to East Mall at UBC, and along the east side of Allison Rd between University Blvd and Dalhousie St; street plantings (mostly mixed with the Common Catalpa, so that the two may be compared) are found along 9th Ave from Vine St to Trafalgar St, on 10th Ave from Stevens St to Blenheim St, on 12th Ave from Trafalgar St to Vine St, on Angus Dr from 33rd Ave to 37th Ave, and on Nelson St from Gilford St to Bidwell St; and there are several large specimens in front of the Canadian National Railways Station between Station St and Main St.

Paulownia tomentosa Steud.
Princess Tree, Paulownia, or Foxglove Tree

This large, deciduous tree from China is cultivated for its large, showy flowers, and is commonly naturalized in the southeastern United States. It is very showy in May when the large blue-purple, tubular flowers appear before the foliage emerges. Over-wintering flower buds with a covering of tan felt are also attractive, but, unfortunately, are frequently killed in our climate. The large leaves are very similar to those of the catalpas. People who have seen the Jacaranda tree in tropical countries often mistake Paulownia for it.

It is not common here, but there are a number of trees around which do flower from time to time, especially after mild winters. There are two in front of the Canadian National Railways Station between Station St and Main St, a large one on the west side of sw Marine Dr just south of 64th Ave, a large one at the north end of the pedestrian walkway by the causeway in Stanley Park, one at the entrance and one in the Sino-Himalayan Garden at Vandusen Botanical Garden, and two by the new lookout at the entrance to UBC Botanical Garden.

Caprifoliaceae – Honeysuckle Family

Sambucus nigra L.
Common Elderberry or Black Elderberry

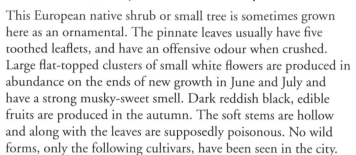

This European native shrub or small tree is sometimes grown here as an ornamental. The pinnate leaves usually have five toothed leaflets, and have an offensive odour when crushed. Large flat-topped clusters of small white flowers are produced in abundance on the ends of new growth in June and July and have a strong musky-sweet smell. Dark reddish black, edible fruits are produced in the autumn. The soft stems are hollow and along with the leaves are supposedly poisonous. No wild forms, only the following cultivars, have been seen in the city.

'Albo-variegata' – This form has leaves edged in white. The only one seen is a tree form on the north side of the Stanley Park Pitch & Putt Golf Course.

'Aurea' – This is the gold-leaved form, and although rare, it is the most commonly cultivated form. There are several tree-like individuals in the city, including one on the NE corner of Laurel St and 13th Ave, one on the north side of 8th Ave just west of Macdonald St, and one on the SE corner of 5th Ave and Maple St.

The common native elderberry in forests in our area is the Red Elderberry (*Sambucus racemosa* L.). It usually remains shrubby, but there are individuals which reach small-tree proportions, especially in the edge of Pacific Spirit Regional Park along Spanish Banks, and in Stanley Park. It is rarely, if ever, cultivated.

Viburnum opulus L.
European Cranberry Bush

The viburnums are a common temperate group of shrubs that are quite ornamental. However, a few species, such as this one, do reach small-tree size. This Eurasian species is usually seen in cultivation as the very double, sterile form 'Roseum,' rather than as the wild, fruiting form. The typical wild tree produces large, flat-topped flower clusters of small, 5-petalled, white flowers surrounded by a ring of much larger, sterile white flowers. The peak of flowering is usually in the middle of May. The fertile flowers produce large clusters of drooping scarlet fruits in the autumn, often remaining on the plants well into winter. The opposite, 3–5-lobed, toothed leaves are similar in shape to those of some of the maples. The leaves turn a dull pink to wine or red in autumn.

Larger wild, single forms in the city include a multi-trunked tree, 5–6 m tall, on the north side of 57th Ave between Cartier St and Hudson St, one on the NW corner of 50th Ave and McKinnon St, and one on the north side of 11th Ave between Sasamat St and Tolmie St.

'Roseum' ('Sterile'), Snowball Bush – This is the cultivar most widely grown and the one most likely to become tree-like. The flowers are all sterile, forming large, round, pendulous balls of flowers, beginning as a pale chartreuse then turning pure white and often fading with a blush of pink. No fruits are produced on this cultivar. Among the larger specimens in the city are a single-trunked tree on the SE corner of 14th Ave and Courtenay St, one on the north side of 12th Ave between Blanca St and Tolmie St, one on the SW corner of 17th Ave and Macdonald St, and three large ones on the SW corner of 46th Ave and Prince Edward St.

Celastraceae – Staff-Tree Family

Euonymus europaeus L.
European Spindle-Tree

Most of the members of this genus are smaller shrubs, but this European native definitely becomes a small tree. It is rather nondescript for most of the year, with green, angular twigs, opposite rounded to oblong leaves, and small, yellow-green, 4-petalled flowers (in June). The 4-angled, reddish pink fruits in late summer and autumn are the showiest feature of the tree. These split to reveal bright orange seeds.

It is a relatively common small tree around parks and gardens in the city. There is a row of four on the east side of Valley Dr between 22nd Ave and 23rd Ave; one on the western side of the Quarry Garden in Queen Elizabeth Park; and several in Stanley Park, including three in the perennial beds, one on the west side of the seal cage at the Zoo, one north of the Stanley Park Dining Pavilion, and two on the south end of the pedestrian overpass by the causeway.

Cercidiphyllaceae – Katsura Tree Family

Cercidiphyllum japonicum Siebold & Zucc.
Katsura Tree

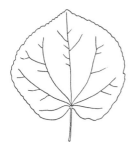

This tree gets its scientific name from the shape of the leaves (phyllum) which look like those of the redbuds (*Cercis*). Leaves are borne both on short, thick side shoots (known as short or spur shoots), and on longer, slender shoots at the ends of branches. Young growth often begins to emerge by late February to early March and is a beautiful coppery pink. It also provides some of our best autumn colour, usually in pastel shades, in late September and early October. The leaves go through colour changes from soft purple to pink, orange and yellow, often all on a single tree at one time. The dull red, petal-less flowers are produced on short shoots before or just as the new leaves emerge, with male and female flowers borne on separate plants. Most of the specimens in Vancouver are quite young, but the Katsura Tree eventually becomes very large with a distinctly pyramidal outline.

male flowers

female flowers

There are street plantings on Selkirk St between 40th Ave and 41st Ave, on Brightwood Place off Vivian Dr (opposite Fraserview Golf Course entrance), in front of the Vancouver School of Theology (west of the Centre for Continuing Education), and south of Chancellor Blvd at UBC; there are several trees in the parking lot of VanDusen Botanical Garden and in the Asian Garden at UBC Botanical Garden, a fairly large one on the NW corner of Panhellenic House, and several in the courtyard of the Biological Sciences Bldg on the UBC campus.

Cercidiphyllum magnificum is a similar species with larger, puckered leaves that is considered by some authorities to be merely a form of *Cercidiphyllum japonicum*. There is also a cultivar with weeping branches named 'Pendula.' There are a few specimens of this species and the cultivar in UBC Botanical Garden and VanDusen Botanical Garden. •31, 32

twig with fruits

Cornaceae – Dogwood Family

Cornus – Dogwoods

This is a variable genus, containing everything from the small herbaceous Bunchberry (*Cornus canadensis*) to tall trees, but the majority of the species are shrubby. Almost all dogwoods have broad, round, untoothed leaves, with a prominent midrib and prominent lateral veins that quickly curve and become almost parallel. Most have opposite leaves. Flowers, usually creamy white, are small and 4-petalled, but grow in tight or loose clusters of a few to hundreds. Most dogwoods have only these small flowers, but some of the more familiar species grown as ornamentals have 4 or more large showy white to pink bracts beneath the flowers. In the bracteate dogwoods, the seeds are surrounded by fleshy red pulp, remaining distinctive at maturity as in the Eastern Dogwood (*Cornus florida*) or becoming fused into a round ball as in the Chinese Dogwood (*Cornus kousa*). The shrubby species and the Great Dogwood (*Cornus controversa*) have seeds covered with white, blue, or black flesh, and the fruits remain separate at maturity.

Cornus leaves

1　*C. controversa*
2　*C.* 'Eddie's White Wonder'
3　*C. florida*
4　*C. florida* 'Rainbow'
5　*C. florida* 'Tricolor'
6　*C. kousa*
7　*C. mas*
8　*C. nuttallii*
9　*C. nuttallii* 'Gold Spot'

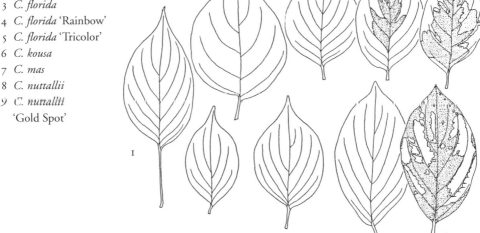

Cornus controversa Hemsl.
Great Dogwood or Giant Dogwood

Mature specimens of up to 20 m tall of this large deciduous tree from China and Japan are spectacular sights, but old specimens in North America are rare. It is said to be the largest of all dogwoods, but some trees of our native *Cornus nuttallii* may equal it. The trees have a very beautiful tiered or layered branching pattern, with flat-topped, creamy white flowers borne in profusion along the tops of the branches in May. The individual flowers are only a few millimetres across, but they are borne in such large clusters that they are attractive from a distance. The flowers are followed by small round fruits that change from green to dark blue. The leaves are typical for a dogwood, with pairs of prominent, parallel lateral veins, but have longer petioles than any of the other local dogwoods. There is a rare variegated-leaved form that is even more showy.

It is extremely rare in the city. There is a large one on the eastern end of the Pitch & Putt Golf Course in Stanley Park, but, unfortunately, this tree has been limbed-up quite high and is difficult to see clearly except from a distance. There are several young specimens in the Asian Garden at UBC, which are now beginning to assume their typical layered look and are flowering well. •33

Cornus 'Eddie's White Wonder'
(*Cornus nuttallii* Aud. × *Cornus florida* L.)
Eddie's White Wonder Dogwood

This is a most interesting hybrid because it is of local origin. The cultivar is the result of crosses made by a pioneer Vancouver nurseryman, Henry M. Eddie, during the late 1930s and early 1940s. The crosses involved both white and pink Eastern Dogwood (*Cornus florida*) and our native Western Dogwood (*Cornus nuttallii*). Most of the seedlings were lost to a flood in 1947 in the Fraser Valley, but one of the best seedlings had been moved to Richmond and this is the one from which all of the 'Eddie's White Wonder' have been propagated. The hybrid is more like the Western Dogwood in many respects, especially in the size of the large bracts, but there are usually only four bracts and they often have a slight indentation at their tips, characteristics of the eastern parent. The tree is sterile and has very small clusters of flower buds, which do not usually open. It flowers a week or two after the peak of flowering of our native dogwood, usually in early to mid-May and lasting for several weeks. The habit is more weeping than either parent. Autumn colour is usually good, with shades of copper, red, and orange. It seems to be more disease-resistant than either parent and is one of the most attractive and prominent hybrids to have arisen in the British Columbia nursery industry.

One of the original seedlings is in the garden of the hybridizer's son on the corner of sw Marine Dr and 41st Ave. There are street plantings along 57th Ave from Arbutus St to sw Marine Dr, four on the se corner of Alder St and 10th Ave, and several around the parking lot at VanDusen Botanical Garden and west of the Pitch & Putt Golf Course in Queen Elizabeth Park. A planting along the south side of 41st Ave from sw Marine Dr to McCleery St is mostly of the hybrid, but there are a few white *Cornus florida* and one large *Cornus nuttallii* at the eastern end, so that both parents and the hybrid may be observed together. •35

Cornus florida L.
Eastern Dogwood or Flowering Dogwood

This very common small tree reaches 10 m tall and is one of the showiest of all spring-flowering trees in the edges of deciduous forests in the East, extending from southern Ontario to Florida and Mexico. It has been adopted as the state flower of several American states. Unlike the Western Dogwood (*Cornus nuttallii*), the Eastern Dogwood characteristically has four showy white bracts surrounding the small yellow-green flowers. The bracts have a distinctive red-purple notch at their tips which is the best characteristic for separating this dogwood from all others. Very broad, wavy leaves have simple edges and are pointed at each end. As with almost all dogwoods, the leaves are borne in opposite pairs along the stems.

Sadly, the trees never look as good in our cool summer/wet winter climate as they do in the East. They are excessively twiggy, the leaves are smaller, as are the flower bracts which often stay caught at the tips when opening. No explanation has been found for this strange condition, but it apparently has to do with our cool, wet spring weather during the time when the bracts are expanding, although it may reflect something about the previous summer's weather when the buds were forming. For example, the bracts expanded much better than usual, during the springs of 1987 and 1988, which followed summers that were long and hot. The peak of flowering is generally about 5–15 May, but the bracts remain showy for a few weeks longer.

The typical wild white form is less common here than the pink forms. Some of the larger white forms in town are on the NW corner of Tecumseh Ave and The Crescent, on the SW corner of 12th Ave and Tolmie St, and a street planting on both sides of 45th Ave from Larch St to Vine St.

'Rainbow' ('Welchii') – This form has a broad yellow-green band around the leaf and good red autumn colour, but usually produces few flowers. There are a number of plants on the south and east side of the BC Place Stadium and along 37th Ave outside VanDusen Botanical Garden; there is one on the north side of 15th Ave between Vine St and Balsam St, one on the east side of Vine St between 49th Ave and 51st Ave, one on the NW corner of 43rd Ave and Carnarvon St, and one by the Main Garden Centre in UBC Botanical Garden. •34

'Rubra,' Pink Flowering Dogwood – This is the more common cultivated form, with dark pink bracts, and is seen here more often than the wild white form. There are relatively large ones on the NE corner of 22nd Ave and Valley Dr, three on the SE corner of McCleery St and 45th Ave, two on 39th Ave between Selkirk St and Hudson St, and some in front of the Main Library on the UBC campus.

'Tricolor' – This is a very attractive and rare cultivar that has pale grey-green leaves with a creamy white margin flushed with pink. It is also a poor flowerer, but the showy leaves make up for the lack of flowers. There is one on the SW corner of 12th Ave and Arbutus St, and one on the south side of 12th Ave between Windsor St and Glen Dr.

Cornus kousa Hance
Kousa Dogwood or Chinese Dogwood

This choice, round-headed, small tree reaches 6 or 7 m at maturity and is a native of Japan. The flowers are held upright on the branches in June, with each cluster of tiny flowers surrounded by four, pointed, creamy white bracts. The fruits are cherry-like, dull red-purple, and held upright on the branches in autumn. Although pulpy, they are edible. There is a great deal of variation in the size and shape of the trees, the width of bracts, and the size and numbers of fruits produced. Some individuals hold their bracts for many weeks, which often turn bright pink before dropping. The tree always attracts attention because it is the latest to flower of our cultivated dogwoods and because it is not well known to the public. It is

just beginning to become a popular tree, so there are a number
of small trees in the city. It deserves to be much more widely
planted.

Notable plantings include specimens in Stanley Park NW of
Lagoon Dr and Haro St and around the Pitch & Putt Golf
Course; street plantings along the west side of Cambie St from
49th Ave to 54th Ave; a long street planting along Charles St
from Nanaimo St to Renfrew St; single-trunked tree forms
along Campus Rd from Wesbrook Cres to Allison Rd; a group
of large specimens on the north side of Kingston Rd just west
of Acadia Rd; a nice pair of specimens in Queen Elizabeth Park
between the conservatory and the rest rooms, and around the
Pitch & Putt Golf Course; and specimens in the Asian Garden
at UBC Botanical Garden, and along the Rhododendron Walk
in VanDusen Botanical Garden. •36, 37, 38

Cornus mas L.
Cornelian Cherry

This Eurasian deciduous shrub or small tree grows to about 7 m
tall, with clusters of small, greenish yellow flowers in late winter
or early spring, followed by cherry-like, dark red, edible fruits
in summer and autumn. Most individuals do not fruit heavily
in our climate, but are very attractive in the autumn in the
years that they do. The plant's major landscape feature is the
very early flowering habit, usually in February to March, a time
when we are searching for signs of spring. The flower clusters
lack the showy bracts found in our other popular arborescent
dogwoods, and the leaves are smaller than those of most of our
other cultivated dogwoods.

It is not very common in the city. Large specimens are
noticeable when in bloom west of the Quarry Garden in Queen
Elizabeth Park, east of the Pitch & Putt Golf Course in Stanley
Park, in the NE corner of VanDusen Botanical Garden, on the
SE corner of 10th Ave and Cambie St (City Hall grounds), on
Western Cres at Kingston Rd, and in the Winter Garden of
UBC Botanical Garden; there is a large single-trunked tree form
along Lower Mall across from the Old Arboretum at UBC, and
there are two on the west side of the Metallurgical Engineering
Bldg at UBC.

Cornus nuttallii Aud.
Western Dogwood, Pacific Dogwood, or Mountain Dogwood

This beautiful tree, well known locally as our official Provincial Flower, is native from British Columbia to California and Idaho. It becomes a much larger tree than the Eastern Dogwood (*Cornus florida*) and its flowers have more bracts than the latter, 5 or usually 6. The bracts, which are rounded or pointed at their tips, begin to show some coloration by late March and are often fully expanded by mid-April and peak at the end of April. The fruit is a round red ball, usually with only a few seeds developing in each fruit. Our native dogwood has longer leaves than most of the other species. The trees have had disease problems in recent years, a leaf blotch causing leaves to turn brown and drop during the summer.

There are street plantings along Cambie St from 33rd to 36th, in the median of King Edward Ave from Willow St to Arbutus St, and in the median of Chancellor Blvd from Acadia Rd to NW Marine Dr; there are good individuals on the NW corner of 45th Ave and Yew St, the NW corner of 3rd Ave and Trutch St, the NW corner of 10th Ave and Birch St, and on the south side of 6th Ave just east of Blanca St. A very large individual in a garden on the south side of 5th Ave between Balsam St and Larch St flowers heavily in late summer and autumn. •39

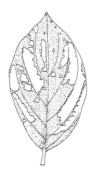

'Gold Spot' ('Eddiei') – This cultivar has irregular gold blotches and streaks on the leaves. It tends to flower well both in the spring and again in the autumn. It was propagated from an individual found near Chilliwack by Henry M. Eddie, a well-known local nurseryman. Specimens around the city include those on the north side of 4th Ave just east of Trimble St, along Cambie St from King Edward Ave to 29th Ave, on the east side of Vine St between 3rd Ave and 4th Ave, and between the parking lot and the entrance to VanDusen Botanical Garden.

Ebenaceae – Ebony Family

Diospyros virginiana L.
Common Persimmon

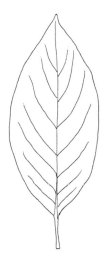

A common eastern North American deciduous tree, Persimmon grows to about 20 m tall. It is rather nondescript in summer with its oval to elliptic, shiny leaves and small green flowers. The leaves often turn orange or red in autumn. The trees may remain showy after the leaves drop, if the orange fruits (2–3 cm across) are produced in quantity. The fruits are full of tannins and are astringent when unripe, but become soft and quite palatable after frosts in the autumn. The bark on older trees is pale grey and breaks into very characteristic, small, square or rectangular plates. It is the hardiest member of a very large genus of mostly tropical trees that includes Ebony, known for its desirable wood. The Japanese Persimmon, *Diospyros kaki,* produces large bright orange fruits to 10 cm long, that have been seen often in local supermarkets in recent years.

The only specimen of Common Persimmon seen locally is a relatively large tree in the Old Arboretum at UBC. Its fruits never ripen here.

Elaeagnaceae – Oleaster Family

Elaeagnus angustifolia L.
Oleaster or Russian Olive

This somewhat willow-like, small tree is the most arborescent member of a genus of mostly large shrubs, often cultivated in our gardens for their silvery foliage and fragrant flowers. It is native from Europe to the Himalayas. The silvery twigs and foliage are attractive throughout the summer. Leaves vary from narrow and willow-like to slightly broader, and are densely covered with flat silvery scales, especially beneath. The foliage is deciduous but some leaves are often held on until late in the winter, especially on young vigorous shoots. Dull yellow, 4-petalled flowers about 1 cm long are produced in spring just after the leaves develop. The flowers are not showy but they may attract attention with their strong sweet fragrance in May. This small tree might be confused with the Willow-Leaved Pear (*Pyrus salicifolius*) when not in flower or fruit, but the pear has stiffer, thicker twigs and is much rarer in cultivation in our area.

It is not very common locally. There are several small tree forms at the Salvation Army Lodge on the NE corner of 58th Ave and Kerr St, three large specimens above the lower lake at VanDusen Botanical Garden, one on the south side of 10th Ave between Granville St and Fir St, a tall one on the NE corner of 37th Ave and Larch St, and a group at the eastern end of the lake in False Creek Park.

Elaeagnus umbellata Thunb.
Russian Olive

Usually a deciduous or partially evergreen shrub, this *Elaeagnus* may become distinctly tree-like, with one or a few trunks. It is native to the Himalayas, China, and Japan. The leaves are very similar to the previous species, but are broadly oval, very silvery beneath, and slightly greener above. The fragrant dull yellow flowers appear in spring and are often followed by large quantities of dull red fruits to 1 cm long, covered with silvery scales.

There are a number of tree forms in the park on the north side of 11th Ave east of Clark Dr.

Ericaceae – Heath Family

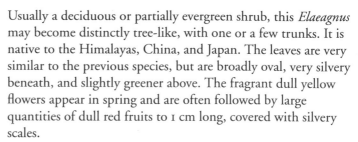

Arbutus menziesii Pursh
Arbutus or Madrone

This is the only native broad-leaved evergreen tree in Canada, reaching its northern limit on rocky bluffs of the southwestern part of our province. A large wild population may be seen along the Upper Levels Highway on the rocky slopes between West Vancouver and Horseshoe Bay. The tree is not often seen in cultivation because of the extreme difficulty in transplanting it. The ones that have managed to grow large are showy throughout the year with their fragrant white flowers, peeling bark, red-orange fruits, and waxy green leaves. Locally, the trees are usually called Arbutus, but they are more commonly called Madrone or Madroño from Washington to California. Those people who are lucky enough to have an Arbutus tree in their gardens do not usually like the trees because of the constant leaf and bark litter.

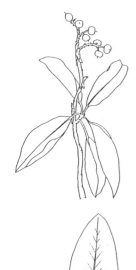

Large specimens around the city include those in McCleery
Park at Marine Cres and 49th Ave, on the SE corner of 57th Ave
and Maple St, on the north side of 5th Ave between Macdonald St and Bayswater St, on the north side of Chancellor Blvd
just east of NW Marine Dr, on the south side of 4th Ave
between Blanca St and Tolmie St, and on the UBC campus
south of MacMillan Bldg. Appropriately, there is a large
specimen on the west side of Arbutus St between 33rd Ave and
34th Ave.

Arbutus unedo L.
Strawberry Tree

This European relative of our native Arbutus is not often cultivated here. The leaves are smaller than those of *Arbutus
menziesii* and the flowers, with their honey-like fragrance, are
produced in late autumn and winter. Attractive edible fruits are
produced in some years and they ripen at the time that the
next year's flowers are in bloom. The fruits are yellow to red,
warty, and strawberry-like, thus the common name. The plant
is usually a dense evergreen shrub but can become tree-like
with very beautiful reddish flaking bark. Unfortunately we
are near their northern limit and they do freeze back, or some
individuals may be killed during our most severe winters.

There are a few plantings around town, mostly shrub-like,
including some on the south side of Granville Island, a tall,
slender tree form just northwest of the Parks Board Office
in Stanley Park among a planting of rhododendrons, and one
on the north side of the Stanley Park Pitch & Putt Golf Course.
The largest one seen in the city is a dense tree form about 3 m
tall on the north side of 10th Ave at Laurel St.

Oxydendrum arboreum (L.) DC.
Sourwood

Sourwood, a small deciduous tree native to eastern North
America, occasionally grows to 20 m tall and is one of our most
desirable small trees for landscapes. It has lance-shaped leaves
and long pendulous clusters of white, fragrant flowers in late
summer, followed by brown seed capsules that are noticeable all
winter as they hang from the ends of the bare branches. The
deep, vertical furrows of the pale grey bark are also evident in
winter. In many years, there is good autumn colour of oranges
and reds very late in the season, although the colour may begin
earlier following dry summers.

It is not very common here, and most of the trees in the city are small. There are three large ones in The Crescent (two near the middle and one near Osler St), one in a garden at Hudson St and 50th Ave, a very large specimen on the sw corner of 29th Ave and Windsor St, two trees (one very large) on the east side of Crown St between 39th and 40th Ave, and several small trees in the Eastern North American Section at VanDusen Botanical Garden.

Rhododendron

Rhododendrons and azaleas are probably our most common landscape shrubs in the city. But, one only has to see the collections around the Stanley Park Pitch & Putt Golf Course, and, especially, the vast collections in the Asian Garden at UBC Botanical Garden or in the Sino-Himalayan Garden and the Rhododendron Walkway at VanDusen Botanical Garden, to realize that many do become distinct trees after a few years. Some species are known to reach 20 m! We do not know how tall many of the modern hybrids may eventually grow because most are still quite young.

Rhododendron ponticum is a species that, together with its hybrids, is commonly grown around the city, and that definitely becomes a picturesque tree with age.

Rhododendron ponticum L.
Ponticum Rhododendron

This late-flowering, vigorous species is often cultivated and becomes tree-like very quickly. It has been used a great deal in hybridization. A native of Spain, Portugal, and Asia Minor, and now widely naturalized in Britain, it eventually becomes an evergreen tree up to 10 m tall. Large flower clusters on the ends of the evergreen branches are produced in late May and June. The individual flowers are pale purple flecked with yellow-brown inside.

The nice old specimens in the city include one on the sw corner of 14th Ave and Birch St, two on the north corner of Chilco St and Nelson St, one on the south side of 12th Ave between Spruce St and Alder St, one on the south side of 36th Ave between MacKenzie St and Carnarvon St, two on the south side of Point Grey Rd between Dunbar St and Collingwood St, one on the sw corner of 29th Ave and Quebec St, and a large one in the centre of The Crescent. •40, 41

Colour Plates

1   *Araucaria araucana* - 33rd Ave
2   *Chamaecyparis lawsoniana* 'Allumii,' 'Erecta
    and *Picea pungens* - City Hall
3   *Chamaecyparis lawsoniana* 'Pendula'- Queer
    Elizabeth Park
4   *Chamaecyparis lawsoniana* 'Stewartii'- 15th /
5   *Ginkgo biloba* - VanDusen Botanical Garde
6   *Abies concolor* - Mackenzie St

7 *Larix × eurolepis* - autumn colour and cones
8 *Picea pungens* and *Acer circinatum* - Queen Elizabeth Park
9 *Picea sitchensis* - cones
10 *Pinus monticola* - Queen Elizabeth Park
11 *Pinus monticola* - cones

13   *Taxus brevifolia* - with fleshy arils
14   *Sequioadendron giganteum* - grove of youn
        trees in VanDusen Botanical Garden
15   *Sequioadendron giganteum* 'Pendula' -
        VanDusen Botanical Garden
16   *Acer circinatum* - autumn colour
17   *Acer davidii* - attractive striped bark
18   *Acer griseum* - autumn colour,
        UBC Botanical Garden

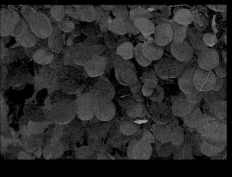

19 *Acer griseum* - flaking bark
20 *Acer macrophyllum* - flowers
21 *Acer palmatum* - grove of trees in autumn
    colour, VanDusen Botanical Garden
22 *Acer rubrum* - autumn colour, Marine Cres
23 *Cotinus coggygria* - autumn colour

24   *Rhus typhina* 'Laciniata' - autumn colour, 7th
25   *Aralia elata* - early autumn colour,
       VanDusen Botanical Garden
26   *Betula papyrifera* - Old Arboretum, UBC
27   *Betula utilis* - peeling white bark
28   *Catalpa bignonioides* 'Aurea' - VanDusen
       Botanical Garden
29   *Catalpa bignonioides* 'Aurea' - flowers

30 *Catalpa* × *hybrida* - flowers
31 *Cercidiphyllum japonicum* - autumn
    colour, Selkirk St
32 *Cercidiphyllum japonicum* - trunk and leaves
33 *Cornus controversa* - flowers
34 *Cornus florida* 'Rainbow'

35   *Cornus* 'Eddie's White Wonder' -

41   *Rhododendron ponticum* - flowers
42   *Gleditsia triacanthos* - fruits
43   *Gleditsia triacanthos* - Queen Elizabeth Park
44   *Laburnum* × *watereri* - UBC
45   *Robinia pseudoacacia* 'Unifoliata' - Queen
     Elizabeth Park
46   *Robinia pseudoacacia* 'Frisia' - VanDusen
     Botanical Garden

47 *Liquidambar styraciflua* - autumn colour, UBC

48 *Liriodendron tulipifera* - autumn colour, Blanca St

49 *Liriodendron tulipifera* - flowers

50 *Magnolia dawsoniana* - VanDusen Botanical Garden

51 *Magnolia dawsoniana* - flower

52 *Magnolia dawsoniana* - flower
53 *Magnolia hypoleuca* - flower
54 *Magnolia kobus* - as street tree on 7th Ave
55 *Magnolia kobus* - ripe fruit exposing
    red seeds
56 *Magnolia sargentiana* - UBC Botanical
    Garden
57 *Magnolia sieboldii* - flower

58 *Chionanthus virginica* - flowers
59 *Fraxinus latifolia* - bark
60 *Fraxinus excelsior* 'Jaspidea' - autumn
   colour, Queen Elizabeth Park
61 *Platanus × acerifolia* - street planting
   on West Broadway
62 *Crataegus × lavallei* - fruits
63 *Crataegus phaenopyrum* - autumn colour
   and fruits

64

65

66

67

68

69

70   *Prunus cerasifera* 'Atropurpurea' and
      Chamaecyparis lawsoniana 'Erecta' -
      street planting on 16th Ave
71   *Prunus padus* - flowers
72   *Prunus serrula* - coppery trunks
73   *Prunus serrula* - flowers
74   *Prunus serrulata* 'Kanzan' - Acadia Rd
75   *Prunus* × *yedoensis* - UBC

76 *Pyrus salicifolia* 'Pendula' - VanDusen
Botanical Garden
77 *Pyrus salicifolia* 'Pendula' - flowers
78 *Pyrus salicifolia* 'Pendula' - fruits
79 *Phellodendron amurense* - autumn colour,
Queen Elizabeth Park
80 *Salix × chrysocoma* and *Robinia pseudoacacia* - early spring in Kitsilano Park
81 *Styrax obassia* - flowers

84

82

85

83

8

82 *Stewartia pseudocamellia* - flower
83 *Stewartia pseudocamellia* - bark
84 *Ulmus americana* - autumn colour, UBC
85 *Ulmus glabra* 'Lutescens' and
    Sequioadendron giganteum - median of
    Cambie St
86 *Zelkova serrata* - autumn colour,
    VanDusen Botanical Garden

Eucommiaceae – Eucommia Family

Eucommia ulmoides D. Oliver
Eucommia

This deciduous tree, growing to 20 m tall, is a native of China and the only member of its family. The leaves are elliptic and toothed, somewhat elm-like, as the name *ulmoides* suggests. The green flowers are insignificant, with male and female flowers borne on separate trees. Female flowers are followed by flattened, elongated samaras resembling those of an elm. The bark has been used medicinally in China, and it is the only hardy temperate tree known to produce rubber latex.

The only two specimens located in Vancouver are a slender tree in the Old Arboretum at UBC and a younger, but larger, one in the Asian Garden of UBC Botanical Garden.

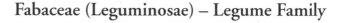

Fabaceae (Leguminosae) – Legume Family

Albizia julibrissin Durazz.
Mimosa or Silk Tree

This is one of our most tropical-looking trees, with its broad, flat-topped growth habit, doubly pinnately-compound leaves with very tiny ultimate leaflets, and showy pale to dark pink 'pom-poms' of very fragrant flowers in late summer and early autumn. The flowers are made up of a large cluster of small, petal-less flowers, but with many long showy stamens. The pale brown pods are like very flat beans. It is native from Iran to central China, but has been cultivated for several centuries in subtropical and warm-temperate parts of the world. It is the only relatively hardy member of a genus of 100–150 species of ornamental tropical trees, many of which are grown in parks and gardens. Mimosa is considered a very valuable and unusual ornamental in our area, as we are at its northern limit for even moderate success, although it reseeds and becomes weedy in warmer parts of the world. It is very commonly naturalized in the southeastern United States.

There are a number of small trees around town, but among the largest are individuals on the east side of an office building on the sw corner of 5th Ave and Alberta St, one in front of an apartment complex on the north side of 4th Ave at Collingwood St, a very attractive dark pink-flowered specimen on the west side of Collingwood St between 1st Ave and 2nd Ave, and one on the north side of 27th Ave just east of St. George St, probably the largest one in the city is on the sw side of 64th Ave west of Columbia St.

Cercis canadensis L.
Eastern Redbud

A common tree or large shrub growing to about 15 m tall, Eastern Redbud is usually found in the edges of the eastern deciduous forest from the southernmost tip of Ontario, New Jersey, and Michigan south to Florida and Texas. It often grows with the Eastern Dogwood, and these two trees produce a spectacular display of colour in the early spring in the East. The leaves are heart-shaped and have a distinctive thickening where the petiole joins the leaf blade. Pea-shaped flowers of bright magenta to purplish pink are borne in profusion along the twigs in late April or early May, before the leaves emerge. Flowers may also be borne along the major limbs and even on the trunk, a condition known as 'cauliflory.' Flat bean-like seed pods, 6–9 cm long, initially green or often bright purple-red and later turning brown, are noticeable in summer and autumn. However, the tree does not usually flower and fruit as well in our area, probably because of our short period of summer heat.

It is an uncommon tree in the city. There is one on the east side of Highbury St between 19th Ave and 20th Ave, four trees west of the Quarry Garden in Queen Elizabeth Park, five trees along the pedestrian walk on the north side of False Creek, two groups of trees at the Place Vanier Residences at UBC, and a grove of young trees in the recently planted Canadian Heritage Garden at VanDusen Botanical Garden.

'Forest Pansy' – A cultivar with dark purple leaves that remain attractive throughout the summer. There is a specimen in the Pine Woods at VanDusen Botanical Garden and there are small ones on the NE corner of Highbury St and King Edward Ave and on the north side of Point Grey Rd just east of Dunbar St.

Cercis siliquastrum L.
Judas Tree or Mediterranean Redbud

Similar to the previous species, this *Cercis* grows in the wild from the Mediterranean to western Asia. It gives a spectacular display against the predominately grey foliage of many of the common Mediterranean plants. The leaves are rounder and less pointed than those of the Eastern Redbud and the flowers and pods are a bit larger. Bright rose pink flowers are borne in profusion along the stems, and sometimes the main limbs or even the trunk in May in Vancouver. The flat pods are usually bright pinkish red and showy during the summer. The brown pods hanging on bare branches over winter are not very attractive.

The only trees seen in Vancouver are a very nice specimen in the lawn just north of the Mediterranean Garden at VanDusen Botanical Garden and one in the garden area NW of the Main Library at UBC.

Cladrastis kentukea (Dum. Cours.) Rudd
(*Cladrastis lutea* [Michx. f.] C. Koch)
Yellowwood

Most books list the species as *Cladrastis lutea,* but it has been shown recently that *kentukea* was an older name, and thus has priority. This leguminous tree from the southeastern United States gets its name from the wood which is bright yellow when first cut. Trees grow to about 12–15 m tall, with very smooth bark. Pinnately-compound leaves are borne alternately along the stems, with 5–7 round leaflets, with the end one usually being the largest. The autumn colour is bright yellow. Individual 2.5 cm flowers are white, fragrant, and borne in large clusters at the tips of branches in early June, although they are not always produced in our climate.

It is very rare in Vancouver. There are two magnificent specimens on the south side of The Crescent, and a young one in the Eastern North American Section at VanDusen Botanical Garden.

Gleditsia triacanthos L.
Honey Locust

A deciduous tree from the eastern United States which reaches as far north as the southernmost tip of Ontario, Honey Locust is usually small but occasionally reaches 40 m in height. The general shape is reminiscent of the related acacia trees of Africa. There are stout, often branched thorns on the limbs and trunk, although many of the cultivated forms are thornless or nearly so. It is very late to leaf out in the spring. The leaves are usually pinnately-compound, with many narrow leaflets, 1–3 cm long, although the leaves may be partially or completely doubly

compound, especially toward the ends of very vigorous shoots. Flowers are borne in long spikes, but are very small and green, and are generally not noticed. The large, flat, twisted pods, 30–45 cm long, hang from the bare branches over winter. It is becoming an increasingly popular street tree here and elsewhere, largely because the small leaves do not cause a litter problem when they fall. There are several selected forms that are vegetatively reproduced.

There are good street plantings on both sides of 10th Ave between Discovery St and Courtenay St (all green forms except one 'Sunburst') and on the south side of 31st Ave from Carnarvon St to Blenheim St; there is a large tree on the NE corner of 4th Ave and Pine St; and there are three trees on the sw corner of 8th Ave and Macdonald St (in front of the library) which usually produce the long seed pods, and several green forms west of the Rose Garden in Queen Elizabeth Park. •42, 43

'Sunburst' – The bright golden yellow foliage of this tree is very distinctive, especially when the leaves first emerge in the spring and at the ends of branches during the summer. It is generally more popular than the green-leaved forms and it leafs out a bit earlier in the spring. There are no thorns, or very few, and seed pods are not produced. There are a number of small trees here, but not many large ones. There is one at the entrance to Stanley Park, five on 2nd Ave west of Yew St, a row on the south side of King Edward Ave from Cambie St east to Yukon St, and a row on the north side of 40th Ave between Windsor St and St. Catherines St.

Gymnocladus dioica (L.) C. Koch
Kentucky Coffee-Tree

This deciduous leguminous tree of the eastern United States and as far north as southern Ontario grows to 35 m tall. The compound leaf is up to 1 m long and is made up of variable numbers and combinations of broad, ovate leaflets. Small greenish white flowers are produced in terminal panicles on separate male and female trees. The female flowers are followed by large (25 cm long by 5 cm wide) dark brown, typically bean-like pods.

It is rare here. There is a specimen in the Old Arboretum at UBC, two small ones in the Eastern North American Section at VanDusen Botanical Garden, and a small one west of the Pitch & Putt Golf Course in Stanley Park.

Laburnum anagyroides Medic.
Common Laburnum or Golden-Chain Tree

This small leguminous tree is a native of southern Europe but is commonly cultivated in gardens for its long chains of golden yellow flowers that appear in May. Leaves are composed of three elliptic to oval leaflets, each about 7 cm long. The pendulous inflorescences vary in length from tree to tree, but are generally 10–15 cm long, about half the length of those of the similar Hybrid Laburnum. Individual flowers are typically pea-shaped and about 2–3 cm long. The combination of golden chains of flowers and the bright green leaves is an unforgettable sight. Flowers are followed by flat pea pods about 5 cm long, usually produced abundantly, and these are very apparent on the trees from midsummer into winter. Unfortunately, all parts of the tree are poisonous, especially the seeds. Common Laburnum is happy enough with our climate to reseed freely and is now naturalized locally in hedgerows, vacant lots, and edges of forests, but is not as commonly cultivated in Vancouver as the Hybrid Laburnum. The trees usually have slender, more sinuous, and, often, multiple trunks when compared to the thick, straight trunks of the Hybrid Laburnum. Common Laburnum flowers ten days to two weeks earlier than the Hybrid Laburnum.

There are many small trees around the city, including one at the south end of the Burrard Bridge, on the east side of Alma St between 11th Ave and 12th Ave, on the north side of 16th Ave just west of Arbutus St (and also just west of Vine St), on the south side of 15th Ave at Alder St, and a number of trees on the NE corner of 10th Ave and Spruce St. There are specimens of both the common and Hybrid Laburnum on the NW corner of 13th Ave and Sasamat St, where the two may be compared.

Laburnum × watereri (Kirchn.) Dipp. (*Laburnum alpinum* [Mill.] Bercht. & J. Presl × *Laburnum anagyroides* Medic.)
Hybrid Laburnum

This garden hybrid combines the best flower characteristics of each parent, with longer chains (20–30 cm long and often with 60–90 flowers) of larger flowers than the Common Laburnum. It is not always easy to distinguish from the latter, especially when in flower, unless the two are growing together. However, the hybrid produces very few pods, each containing only one or two seeds, compared with the large numbers of pods, usually

with a number of seeds, produced by the Common Laburnum. The hybrid is also usually grafted on a single, straight trunk of Common Laburnum, so that the shape of the tree looks different from the common form, which grows on its own trunk. The other parent, the Scotch Laburnum (*Laburnum alpinum*) has not been seen in Vancouver.

Among the largest and most beautiful specimens are those around the UBC campus, including a row of six old trees along Agronomy Rd south of MacMillan Bldg (growing with one Common Laburnum, which is probably the result of sprouts from below the graft), a row of four SW of the Main Library, four on the north side of the Old Armoury (West Mall and Crescent Rd), three old specimens along the west side of Wesbrook Mall just north of Agronomy Rd, and several nice specimens around the parking lot of BC Research on Wesbrook Mall south of 16th Ave. There is a beautiful row of young trees in the Laburnum Walk at VanDusen Botanical Garden, two trees on the east side of Marguerite St between 32nd Ave and 33rd Ave, and one growing with the Common Laburnum on the NW corner of 13th Ave and Sasamat St. •44

+ *Laburnocytisus adamii* (Poit.) C.K. Schneid. (*Cytisus purpureus* Scop. + Laburnum *anagyroides* Medic.) Adam's Laburnum

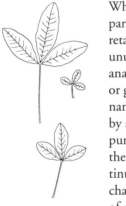

When most plants are grafted onto a related plant, the grafted part (scion) and the one onto which it is grafted (stock) usually retain their separate and distinct characters. A much more unusual situation occurs when the two plants intermingle anatomically, at least partially. This is known as a graft hybrid or graft chimera, and is indicated by a plus sign before the name, so that it is not confused with a hybrid cross (indicated by a times sign). A graft hybrid results when the low-spreading, purple-flowered shrub *Cytisus purpureus* is grafted on limbs of the common tree *Laburnum anagyroides*. The tree then continues to produce most limbs with typical foliage and long chains of yellow *Laburnum* flowers, plus a few rounded clumps of dense twiggy growth with purple flowers, typical of the broom, scattered in the tree, and yet other branches with intermediate leaves and short racemes of dull flesh-coloured or purplish flowers. The intermediate branches do not usually produce seed pods, but the other two types of growth continue to produce their typical pods. The tree is certainly more of a curiosity in botanical collections than particularly ornamental.

There is a single tree along the south side of 41st Ave just west of Hudson St, and there are several near the Laburnum Walk and Rose Garden in VanDusen Botanical Garden.

Robinia pseudoacacia L.
Black Locust or Acacia

This is a deciduous tree from the eastern United States, where it usually grows in the edges of forests and in open fields. It is grown as a park and garden tree in many temperate parts of the world, and is now escaped from cultivation and established widely in its wild form in our area. There are a number of more desirable cultivated forms also grown here. The deeply furrowed bark, dark green pinnately-compound leaves with nearly round or oblong leaflets, and drooping racemes of very fragrant white flowers in early June are all distinctive characteristics of this tree. There is a pair of short, stiff spines beside the swollen base of each leaf. Dry, flat, bean-like seed pods hang on the branches over winter.

There are many specimens of the typical wild form in the city, including a number of trees along 37th Ave (by VanDusen Botanical Garden) from Granville St to Oak St, a row on the west side of Oak St from 37th Ave to 41st Ave, several old specimens on the sw corner of 7th Ave and Pine St, and a nice row of old specimens on the west side of Arbutus St at the western end of Creelman Ave in Kitsilano Beach Park (where the wild form is planted with some of the cultivars listed below). •80

'Decaisneana' – This cultivar is typical of the wild form except that the flowers are a soft pink. There is one tree on the west side of the Rose Garden in Queen Elizabeth Park.

'Frisia' – There are few golden-leaved trees as brilliant as this one. It is just becoming popular locally as a street and park tree. The chartreuse leaves hold their colour from when they first emerge in the spring until autumn when they turn a golden yellow. There is a beautiful pair on the western side of the perennial beds and some small ones around the parking lot at VanDusen Botanical Garden, a small one on the western side of the Law Bldg at UBC, one on the west side of sw Marine Dr just north of 41st Ave, four on Yew St between 3rd Ave and 4th Ave, and a street planting on both sides of 8th Ave from Scotia St to Prince Edward St. •46

'Tortuosa' – Probably the most picturesque and, unfortunately, rare locally. This cultivar has sinuous twigs and small drooping leaves. There are two specimens in Kitsilano Beach Park (one in a row of Black Locust on the west side of Arbutus St at Creelman Ave and one just north of the bathhouse just off the beach), and young ones at VanDusen Botanical Garden near the entrance (just off the deck) and in the perennial beds.

'Umbraculifera,' Mop-Head Acacia – This cultivar lacks thorns and, as the name indicates, forms a round-headed crown of dense small twigs. It is distinctive either in summer when in leaf or in winter when the branch structure is apparent. There is a group of three in the centre of the intersection of University Blvd and Main Mall at UBC (planted about 1931), two on the north side of 24th Ave between Highbury St and Dunbar St, a nice row of five trees along the east side of Hudson St from Dogwood Ave to 52nd Ave, one on the sw corner of 8th Ave and Tolmie St, a row of four on the west side of Dumfries St between 28th Ave and 30th Ave, and an old one on the north side of the Stanley Park Pitch & Putt Golf Course.

'Unifoliata' ('Monophylla Fastigiata') – This form tends to grow very tall and slender, with unusual leaves that have only one large leaflet or a large terminal, plus one or two small lateral leaflets. There is a large tree west of the Rose Garden on the west side of Queen Elizabeth Park, one in the lawn west of the perennial beds at VanDusen Botanical Garden, and one on the west side of Arbutus St at Creelman Ave at Kitsilano Beach Park (with other *Robinia* cultivars). •45

Robinia viscosa Venten.
Clammy Locust

This small locust, reaching 12 m tall, is native to the southeastern United States and is not often cultivated this far north. It is easily distinguished from the Black Locust by the very pubescent stems and rachides of the leaves, especially noticeable on young vigorous growth. The red hairs are very sticky to the touch, thus the common name. The leaflets are larger and rounder than those of the Black Locust. Purple-pink flowers in pendulous racemes are produced in late May and early June. In our relatively cool climate, the trees will probably always remain rather small.

There are two small trees on the se corner of 5th Ave and Alberta St, and one on the east side of Yew St between York St and 1st Ave.

Sophora japonica L.
Japanese Pagoda Tree or Chinese Scholar Tree

This leguminous tree is native to China and Korea and is often cultivated in temperate parts of the world, where it becomes a large picturesque tree. It has smooth bark and dark green, pinnately-compound leaves, made up of 7–17 leaflets. Creamy white flowers in large panicles appear at the ends of the branches in August and into early September. These are followed by green fruits that are like rounded, fleshy beanpods, but they are not often produced in our climate. The bark, wood, and fruits have been used as a source of yellow dye.

This is not a very common tree in the city. There is a row on the south side of 12th Ave just east of Arbutus St, one in front (east side) of the Service Yard Bldg near the Rose Garden in Stanley Park, one in the Old Arboretum at UBC, relatively large specimens at the western end of the Rhododendron Walkway and above the lower lake at VanDusen Botanical Garden, one in the median strip of Cambie St at SE Marine Dr, and a pair at the Dunbar Community Centre on the SW corner of Dunbar St and 31st Ave.

Fagaceae – Beech Family

Castanea sativa Mill.
Spanish Chestnut or Sweet Chestnut

The common edible chestnut of southern Europe, western Asia, and northern Africa is a large tree reaching a height of 35 m or more. The large lance-shaped leaves have prominent, forward-pointing teeth. They are pubescent beneath, especially when young, and are slightly heart-shaped at the base. Typical of all members of the genus, the female flowers are rounded, green, and spiny and the males are long, slender, creamy white catkins with a very foul smell. The June to early July flowers are followed by large spiny burs surrounding one or more edible nuts. Sometimes the trees have moderately good, yellow autumn colour, and some brown leaves often hang on the trees late into winter.

There are a few large trees in the city, usually bearing nuts each year. There is a row of very large ones as a street planting along 20th Ave between Cypress St and East Blvd, one on the south corner of Jervis St and Barclay St, one in a garden on the north side of Chancellor Blvd just west of Blanca St, one in Stanley Park north of the Rose Garden, and a large tree on the east side of Queen Elizabeth Park just below the road.

The American Chestnut (*Castanea dentata*) was once a major deciduous tree in the eastern North American forests but is now nearly extinct in the East due to the Chestnut Blight. There are reports of it being cultivated in our area, but no definitely identified specimens have been found. The leaves are narrower, especially at the base, and are not pubescent beneath.

Fagus – Beeches

There are about ten species of these majestic trees in the northern temperate regions of the world. They are easily identified by their very smooth bark, even on old trees, thick rounded to oval leaves, and pairs of triangular nuts surrounded by soft prickles. There are three species cultivated here but only one, the European Beech (*Fagus sylvatica*), is common and this is extremely variable in its growth habit, leaf shape, and colour. The other two species are surprisingly uniform in characteristics from tree to tree. Many of the cultivars of European Beech may be seen in Queen Elizabeth Park, Stanley Park, and, especially, in a very good collection at VanDusen Botanical Garden.

Fagus grandifolia J.F. Ehrh.
American Beech

This is a dominant deciduous tree in eastern North America, forming vast stands, usually with maples, in what are referred to as beech-maple forests. The smooth grey bark looks very silvery, especially in winter light. The wide spreading trees grow to as much as 30 m tall. The leaves have 9–15 pairs of veins, distinctive marginal teeth, and are generally longer than the much more commonly cultivated European Beech. Many of the leaves remain golden brown and hang on the trees well into winter.

It is a rarely cultivated tree here. The only specimen known in the city is one on the south side of the Stanley Park Pitch & Putt Golf Course.

Fagus orientalis Lipsky
Oriental Beech

This eastern European beech grows to about 35 m tall in its native habitat. Its leaves are long and generally shaped like those of the American Beech, but there are no teeth along the

margins and there are 7–10 pairs of veins. Leaves tend to be slightly thicker and darker green than those of the American Beech.

It is rare here. There is a grove of specimens (with several different forms of European Beech) on the hill on the west side of Queen Elizabeth Park, and one in the Beech Collection at VanDusen Botanical Garden.

Fagus sylvatica L.
European Beech

This massive deciduous tree grows to about 30 m tall and has a broad, wide spreading form when grown in the open. There are many selected cultivated forms with a columnar or weeping habit and coloured leaves, especially purple or coppery. The tree is common in Europe and is widely cultivated as a park or garden tree, mostly in larger gardens as it becomes too massive for the average garden. The bark is pale grey and very smooth even on old trees. Leaves are nearly rounded in outline, with untoothed margins and 5–9 pairs of distinctive veins. Young leaves in the springtime have is a fringe of white hairs around the edges. In summer the leaves on typical wild trees are pale to bright green, becoming quite thick and papery when mature. They often turn coppery and remain on the twigs late into the winter. There are not very many really massive specimens around the city, but there are enough large ones to give an idea of the mature size.

The typical green form includes specimens between the perennial garden and the Service Yard in Stanley Park, in the middle of The Crescent in Shaughnessy, on the SE corner of Tatlow Park (Point Grey Rd and 3rd Ave and west of Macdonald St), street plantings on 40th Ave between Ontario St and Main St, and on 14th Ave between Cambie St and Yukon St.

'Atropunicea' ('Atropurpurea'), **Copper Beech** – Purple- or bronze-leaved forms are the most popular cultivars. There are nice specimens on 13th Ave from Dunbar St to Waterloo St (alternating with Paper Birches, *Betula papyrifera*), on 13th Ave from Macdonald St to Stephens St, between the perennial garden and the Service Yard in Stanley Park, and in The Crescent in Shaughnessy.

'Dawyckii' ('Dawyck,' 'Fastigiata') – This very stiff columnar form may be seen behind (north of) the War Memorial Gym at UBC, alternating with a similar but slightly less columnar form

'Heterophylla'

'Rohanii'

'Roseo-marginata'

'Rotundifolia'

of English Oak (*Quercus robur* 'Fastigiata'). There are street plantings along Earles St from Kingsway to 38th Ave, and along 1st Ave from Victoria Dr to Lakewood Dr.

'Heterophylla' – This is a very narrow-leaved form with irregular dissections to the leaves. There are nice specimens on the NE corner of Blanca St and 5th Ave, and between the perennial garden and the Service Yard in Stanley Park.

'Pendula,' Weeping Beech – This is a large, green-leaved, weeping cultivar with very pendulous small branches. The trunk begins straight, but the major limbs are very sinuous, often arching out at different angles, giving a very open, irregular appearance to the trees. There are a number of specimens in gardens locally, including a row along the SE side of Chilco St between Barclay St and Nelson St, one at the entrance to Sunset Nursery on 41st Ave at Sophia St, one on Knight St north of 49th Ave, one on the north side of Linden Rd near Quilchena Cres, one just east of the Service Yard in Stanley Park, two on 32nd Ave just east of Granville St, one on the SW corner of MacKenzie St and 36th Ave, and several on Oak St and 37th Ave at the entrance to VanDusen Botanical Garden.

'Rohanii' – A purple-leaved form with irregularly dissected leaves. There are specimens in the groves of beeches on both the east and west sides of Queen Elizabeth Park, and in the Beech Collection at VanDusen Botanical Garden.

'Roseo-marginata' – A very attractive cultivar with purple leaves irregularly margined with pink to white. It is best viewed from beneath, so that the light shines through the pale margins. It is often listed as the cultivar 'Tricolor,' but this is probably a different, much rarer, cultivar. There are two specimens on the east side of Queen Elizabeth Park, one on the NE corner of Maple St and 17th Ave, and one in the Beech Collection at VanDusen Botanical Garden.

'Rotundifolia' – A very slender, irregularly branched tree with the smallest and, as the name indicates, the roundest leaves of any of our beech cultivars. The only ones found in the city are four specimens along the NW side of Hornby St between Pacific Ave and Drake St, two on the opposite side of Hornby St between Davie St and Helmcken St, and one in the traffic circle at the entrance to Totem Park Residences at UBC.

'Zlatia,' Golden Beech – This form has golden, arching, or slightly pendulous new growth, and very large leaves. There is one on the east side of Queen Elizabeth Park and one in the Beech Collection at VanDusen Botanical Garden.

Lithocarpus densiflorus (Hook. & Arn.) Rehd.
Tanbark Oak

This oak relative is found in nature from southern Oregon to California. It is usually a rounded shrub but can become a small tree, sometimes reaching 25 m tall. Its leaves more closely resemble those of a small chestnut (*Castanea*) than the leaves of most oaks. The lanceolate, evergreen leaves, to about 10 cm long, are very thick and have distinctive teeth. They are dull grey-green above and pale and fuzzy beneath. Male flowers are in the form of long chestnut-like, white catkins, with one or more female flowers at the base. The fruits are very much like those of the oaks, with a fringe of stiff, thick hairs on the cup. Vancouver is very near the northern limit for even marginal growth of this common Californian plant.

There is a small, sparse tree on the SE corner of the Pitch & Putt Golf Course in Stanley Park, and a larger, slender tree near the beginning of the Rhododendron Walk in VanDusen Botanical Garden.

Nothofagus antarctica (G. Forst.) Ørst.
Antarctic Beech

This uncommon deciduous tree grows to 35 m tall in nature in Chile and Argentina but is only seen locally as a young, small tree. The small leaves (to 2.5 cm long) are irregularly toothed and have oblique bases, much like a small elm leaf. The growth habit tends to be very irregular, with sinuous, wiry twigs. Scaly, angular nutlets less than 1 cm long are often produced in large quantities. As the trees mature, the bark develops attractive irregular plates of various shades of brown.

It is more of a novelty than actually handsome, so there are few cultivated here. There are a number of relatively large ones in the Southern Hemisphere Section of VanDusen Botanical Garden, one at Place Vanier Residences at UBC, a small one NW of the Stanley Park Dining Pavilion, and a group in Queen Elizabeth Park on the north side of the hill above 29th Ave and in the median strip of Cambie St south of Midlothian Ave.

Nothofagus obliqua (Mirb.) Ørst.
Roble Beech

A deciduous tree from Chile, growing to 35 m tall, Roble Beech is more vigorous, faster growing, and of a denser growth habit than the Antarctic Beech, and the leaves are about twice as large. As the species name indicates, the base of the leaves is

strongly oblique, with the general appearance of a small, irregular elm leaf. The leaves turn a dull yellow in the autumn. There are several other *Nothofagus* species in Australia and New Zealand, but, unfortunately, these are not hardy this far north.

It is even rarer than the Antarctic Beech. There is a large, healthy, 3-trunked tree about 12 m tall in front of the Agriculture Canada Research Station on NW Marine Dr at UBC, and small ones in the Southern Hemisphere Section at VanDusen Botanical Garden.

Quercus – Oaks

In northern climates, oaks are usually thought of as large, deciduous trees with the very familiar pinnately-lobed leaves. However, there are many smaller shrubby oaks, evergreen species, and ones with unlobed leaves or very narrow, willow-like leaves. They all produce the familiar fruit, a large nut inside a shallow cup. The leaves of most oaks are relatively hard, leathery, and full of tannins, thus causing them to decompose very slowly. They often hang on the trees late into the winter, or even until spring when the young leaf buds begin to expand. Coppery brown leaves hanging on trees in winter are characteristic of the oaks.

There are two major groups of oaks grown in our gardens. The red (or black) group have soft, needle-like points on the tips of the lobes and they are generally better adapted to our local conditions. The other group, the white oaks, have rounded lobes and many are from hotter, drier climates than ours, so they usually do not grow as well here. The exception is the commonly cultivated English Oak (*Quercus robur*).

Quercus acutissima Carruth.
Oriental Chestnut Oak

This late-deciduous oak grows to 20 m tall, often holds its leaves well into winter, and is a native of Korea, Japan, and China. The thick, shiny, oblong-lanceolate leaves are 12–18 cm long with rows of bristle teeth along the edges, resembling those of the edible chestnuts (*Castanea*). Acorns, when produced, have their cups fringed with soft spines, similar to those of the North American Bur Oak.

It is rare here. A number of trees were planted in 1985 (for Expo '86) along Pacific Blvd from Davie St to Cambie St and from Smythe St to Beatty St.

Quercus cerris L.
Turkey Oak

A tree similar in general shape and character to the common English Oak (*Quercus robur*), but the Turkey Oak is usually a bit narrower in outline and the lobes of the leaves are much more pointed. The leaves are generally oblong, but are variable in shape and size. The bases are rounded and the lobes end in dull points, lacking the elongated pin-like points of the red oaks. They are dull green and smooth above, and grey-green and soft beneath. Winter bud scales have distinct elongated teeth. The acorns are also very distinctive, the cups being covered with long, mossy scales. The tree grows to about 40 m tall in the wild in southern Europe and western Asia, but it is usually much smaller in cultivation, at least in our area.

It is very rare here. There is a large one in Stanley Park in front of the Parks Board Office (on the bank down toward the water), and a young one in the Oak Collection at VanDusen Botanical Garden. Sadly, a large one in a yard at Park Dr and Cartier St was cut down a few years ago, as was a smaller one at Burrard St and Hastings St.

Quercus coccinea Muenchh.
Scarlet Oak

This is one of the brightest of oaks when in red autumn colour in the eastern American deciduous forest, thus the common name Scarlet Oak. Shiny, deeply 7–9-lobed leaves have sharp bristle tips. Easily confused with Pin Oak (*Quercus palustris*) in cultivation, Scarlet Oak has larger, broader leaves, with deeper sinuses between the lobes, and the growth habit of the two trees is different. Scarlet Oak is cultivated much less often in our area.

There is a large tree in the Old Arboretum at UBC, a relatively large one among other oaks on the west side of Queen Elizabeth Park along Cambie St (growing with the similar Pin Oak, so the two may be compared), a large tree hanging over the west side of Acadia Rd at Kingston Rd (north of Chancellor Blvd), and one on the east side of East Mall just south of University Blvd at UBC.

Quercus garryana Dougl.
Garry Oak

Our only native British Columbian oak, Garry Oak is common on the Gulf Islands and the drier parts of southern Vancouver

Island, and always grows on rocky, well-drained soils. The leaves are thicker, darker, and shinier than those of the similar, very common, English Oak (*Quercus robur*). Old trees have grey bark and characteristic gnarled trunks and limbs, making it one of our most picturesque native trees. It is usually difficult to get established and is generally not very much at home in slightly wetter Vancouver. It is usually called Oregon White Oak, south of the border.

The largest ones in the city include one on the west side of Dunbar St between 19th Ave and 20th Ave, a pair on the north side of 71st Ave west of Granville St, one on the north side of 2nd Ave between Alma St and Dunbar St, and smaller specimens on the north side of Point Grey Rd at the foot of Collingwood St and along a path on the NE side of Queen Elizabeth Park.

Quercus imbricaria Michx.
Shingle Oak

One of the un-oak-like oaks, similar to Willow Oak but with broader leaves, this is another of the eastern North American oaks that is not often cultivated on the Pacific Coast. It grows to 20–30 m tall in its native habitat from Pennsylvania and Michigan to Georgia and Arkansas. Oblong leaves, 10–15 cm long and 3–6 cm wide, have no lobes or teeth except for a single bristle tip at the end. Leaves are thick, with a pale, soft pubescence beneath.

Trees were planted in 1985 (for Expo '86) along Pacific Blvd from Richards St to Homer St.

Quercus lobata Nee
Valley Oak or California White Oak

This is a common large oak of the foothills of interior California, but is very rarely seen this far north. The leaves are very irregularly lobed and vary considerably in size and shape on a single tree and on different trees. The acorns are very long and slender, but they do not seem to be produced here.

There is a fairly large tree on the corner of Nelson St and Park Lane in Stanley Park, although the shape is not typical of those seen in the wild.

Quercus macranthera Fisch. & Mey.
Caucasian Oak

Very thick twigs covered with soft down are a distinctive feature of this large-leaved oak. It is a native of western Asia, especially in the Caucasus Mountains, where it reaches 20 m in height. The leaves are often as much as 15–20 cm long and are shallowly lobed. The specific name means large-flowered, named after the relatively showy male catkins borne in spring.

It is not a very common tree in Vancouver. A few good specimens include a row of three on the north side of the old Health Sciences Bldg along University Blvd just east of East Mall at UBC, several young ones in the Asian Garden of UBC Botanical Garden, two large trees (with a number of other oaks) on the west side of Queen Elizabeth Park along Cambie St, and one on the east side of the hill by the Pitch & Putt Golf Course in Queen Elizabeth Park.

Quercus macrocarpa Michx.
Bur Oak, Bur-Cupped Oak, or Large-Fruited Oak

The name *macrocarpa* means large-fruited, and this eastern North American oak has very large acorns (to 4 cm) that are among the largest of any oak. However, most individuals bear smaller acorns. The acorns are very distinctive, not only in their size but also because of the characteristic long fringe on the edge of the cup. Unfortunately, acorns are not very often seen in our climate. The leaves are extremely variable from tree to tree, or even from one leaf to the next on a single branch. The most typical form has a deep sinus on each side about midway and a large, irregular end lobe. The tree does not do well in our climate, although there are a few decent specimens around, and young trees seem to be very slow to become established here. It is a much more familiar tree on the eastern prairies and eastward.

Probably the largest tree in the city is on the east side of Fraser St between 12th Ave and 13th Ave, but there are also two young, healthy trees in the lawn area of a small park on the NW corner of 10th Ave and Heather St, and a grove of five good specimens on the west side of Queen Elizabeth Park along Cambie St just north of 33rd Ave.

Quercus palustris Muenchh.
Pin Oak

A large deciduous oak of the eastern deciduous forest occurring from south southern Ontario to North Carolina and west to eastern Kansas, but most common in Ohio, Indiana, and Illinois. It is tolerant of heavy, wet soils and is now very commonly cultivated as a street and park tree here. The trees have a very characteristic shape, especially when young, with the lower limbs drooping, the middle ones extending nearly horizontally, and the upper ones pointing upward. The common name comes from hair-points at the ends of the leaf lobes. The leaves have large, round sinuses between the lobes. Leaves are narrower than the less commonly planted (at least here) Scarlet Oak. There is sometimes fairly good autumn colour, with shades of reds and oranges, and the brown leaves are then held on the trees well into the winter.

It is moderately common here as a street or shade tree. Among the plantings are trees along West Blvd between 42nd Ave and 49th Ave, several in front of the Canadian National Railways Station between Station St and Main St, and a grove of smaller ones on the west side of the Student Union Bldg on East Mall at UBC. Probably the largest trees are found along the west side of Cambie St between 20th Ave and 21st Ave. Many of these specimens have been limbed-up and lack the characteristic drooping lower limbs.

Quercus petraea Liebl.
Durmast Oak or Sessile Oak

Easily confused with our native Garry Oak and the English Oak, this large deciduous tree grows to about 40 m tall but is rarely cultivated in Vancouver. The growth habit is slightly narrower than that of the English Oak, and the acorns are sessile (attached directly to the twig), not on a slender stalk as they are on the English Oak. The leaves are deeply lobed, usually with three rounded lobes on each side of the leaf. They are dark, dull green above, with longer petioles than those of the English Oak. The Garry Oak has thicker, very shiny leaves.

There is a very large specimen in the Old Arboretum at UBC, and one in the Oak Collection along Cambie St on the west side of Queen Elizabeth Park.

Quercus phellos L.
Willow Oak

A large and very common tree in the southeastern United States, this is a very un-oak-like oak. The leaves are long and slender (6–10 cm long and up to 2 cm wide) and look much more like those of willow than typical oak, hence the common name. However, the leaves are thick and hard, typical of the oaks. They remain partially evergreen in their native area, but are completely deciduous here. The trees bear typical acorns.

There are street plantings along Pacific Blvd from Homer St to Davie St, from Smythe St to BC Place Stadium, and from Nelson St to BC Place Stadium; there are several relatively large individuals in the Oak Collection at VanDusen Botanical Garden, and a small one by one of the lakes in Vanier Park, near the Vancouver Museum.

Quercus prinus L.
Chestnut Oak

This large, common tree of the eastern deciduous forests has thick leaves with many shallow lobes, resembling the leaves of the true chestnuts (*Castanea*) more than most other North American oaks, thus the common name. However, the lobes are rounded rather than pointed as they are in the chestnuts.

The tree is known in Vancouver only from one large specimen in a row of old oaks in the Old Arboretum at UBC, and another one with a number of other oaks along Cambie St on the west side of Queen Elizabeth Park.

Quercus robur L.
English Oak, Pedunculate Oak

This is, by far, the most commonly cultivated oak in Vancouver. It has relatively small, thin, dull leaves with irregular, rounded lobes. The petiole is very short, and the base of the leaf flares broadly along the petiole. The acorns are large and produced on long stalks, unique among our cultivated oaks. There is often more than one acorn produced on one stalk. In its native habitat of Europe, English Oak may become a large tree, growing to 50 m tall, but there are none quite that tall in Vancouver. It is happy enough here to reseed and is now found as a part of our naturalized vegetation in the University Endowment Lands, Pacific Spirit Regional Park, Stanley Park, and other natural areas. It is often mistaken for the native coastal Garry Oak, which has much shinier leaves, sessile acorns, and is very rare in Vancouver.

The many street plantings of the typical wild forms include those along 13th Ave from Trutch St to Carnarvon St, along Trutch St from Broadway to 15th Ave, along the west side of Dunbar St from 31st Ave to 33rd Ave, and behind the Main Library along East Mall at UBC.

'Concordia'–This beautiful form has bright gold leaves and is a spectacular sight when old. The only specimen seen in the city is a small one above the seawall in Stanley Park east of Lions Gate Bridge and just west of the Georgia St exit.

'Fastigiata' – A fairly common cultivar here, this form has typical leaves, but the trees form broad columns with erect limbs branching near the ground. There is a row by the parking lot north of the War Memorial Gymnasium at UBC (alternating with a similar looking, upright cultivar of *Fagus sylvatica*), a group on the south side of Kerrisdale Park off 41st Ave and Maple St, a long row on the east side of John Hendry Park (containing both the narrowest fastigiated form which holds its dry, brown leaves throughout the winter and a broader form that is usually nearly leafless during the winter), and a group on the western side of Queen Elizabeth Park along Cambie St.

'Fastigiata Cucullata' – Similar to the above form but the leaves are very shallowly and irregularly lobed, or sometimes unlobed, and tend to be cup-shaped. There are a number in Queen Elizabeth Park, including a grove on the western rim of the Quarry Garden, one on the eastern side of the park, and two along the main road (parallel to Cambie St) on the western side. There is one on the western end of a row of *Quercus robur* 'Fastigiata' and *Fagus sylvatica* 'Dawyckii' on the north side of War Memorial Gymnasium at UBC.

Quercus rubra L.
Red Oak or Northern Red Oak

This is one of the most common of eastern North American oaks growing to 30 m tall in the mixed oak-hickory forests. In cultivation it is often wide spreading, forming a broad oval in outline. The uniform, shallowly lobed leaves are characteristic. Young emerging leaves in spring are often pale green, turning very dark green for the summer and dull red or orange-brown in autumn. This is the commonly cultivated oak with bristle-tipped leaves in Vancouver parks and gardens.

There are many beautiful old specimens around the city. One of the largest trees is just above English Bay in the park at the corner of Beach Ave and Bidwell St (plus several smaller Red Oak trees in the park); there is a large individual on the east side of Pine Cres between 29th Ave and 30th Ave; and there are long street plantings along Blenheim St from King Edward Ave to 37th Ave, along 32nd Ave from Wallace St to Camosun St, and a double row along Main Mall at UBC that was planted in 1931.

'Lutea' – This unusual form might be called the Yellow Red Oak! Young leaves in spring are a very bright golden yellow, changing to yellow-green and later green as they mature during the summer. There are specimens on the western side of the Pitch & Putt Golf Course in Stanley Park, one on the west side of Queen Elizabeth Park along Cambie St, and one near the beginning of the Rhododendron Walk at VanDusen Botanical Garden.

Quercus stellata Wangenh.
Post Oak

The species name *stellata* comes from the pubescent undersides of the leaves, which can be seen to be covered with many star-shaped (stellate) hairs when examined under a microscope. These hairs are so thick that the leaf feels soft to the touch beneath. This eastern North American tree is one of the white oak group with broad, rounded lobes, the middle ones on each side being characteristically shallowly lobed. The brown leaves are often held on long into winter.

The only specimen seen in the city is a large one in the row of oaks in the Old Arboretum at UBC.

Quercus × *turneri* Willd.
Turner's Oak

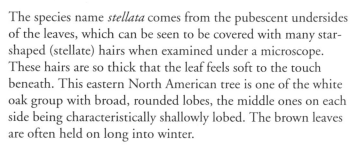

A hybrid between the deciduous English Oak (*Quercus robur*) and the relatively tender, evergreen Holly Oak (*Quercus ilex*), this round-headed tree loses some of its leaves in the autumn, but some leaves usually remain green and hang on the branches throughout the winter. The leaves are dark green, variable in size and shape, but usually very slightly lobed or wavy-margined.

There are a number of trees in our area, including ones along Burrard St at Robson St in front of the Vancouver Public Library, in the Oak Collection in Queen Elizabeth Park along

Cambie St, in front of the Kerrisdale Library on the west side of West Blvd at 43rd Ave, and in the parking lot by the Heather Pavilion at Vancouver General Hospital; probably the largest one in the city is at the north end of the parking lot for the Stanley Park Dining Pavilion.

Quercus velutina Lam.
Black Oak

This is another very large, common oak of the eastern North American deciduous forests. It is common from Maine and southern Ontario to Florida and Texas.

The leaves are long-petioled and have 7–9 broad lobes with bristle-tipped points. The acorns are small. The inner bark has been used as a source of a yellow dye, quercitron.

It is very rare here. The only one known is a large specimen in the Old Arboretum at UBC.

Hamamelidaceae – Witch Hazel Family

Hamamelis – Witch Hazels

These large shrubs or small trees have rounded to oblong leaves with irregular, rounded lobes or teeth, especially near the blunt tips. The trees are generally nondescript during the summer, but their autumn leaf colour is often showy. The trees are at their best in winter, when they are in bloom at a time when little else is in flower. The fragrant flowers have four long, strap-shaped petals of bright yellow, orange, or red. Usually only a few woody seed capsules are produced. These burst with great force, sending the shiny black seeds for long distances.

In addition to the following species, the Hybrid Witch Hazel (*Hamamelis* × *intermedia*), is also widely cultivated. It is a cross between Chinese Witch Hazel (*Hamamelis mollis*) and the Japanese Witch Hazel (*Hamamelis japonica*) and is usually seen as young, shrubbier plants, although some specimens will obviously become tree-like in time. Its flowers vary from orange to copper or dull red.

Hamamelis mollis D. Oliver
Chinese Witch Hazel

This large shrub or small tree grows to 10 m tall in its native habitat in western China, with a typically open, vase-shaped habit. The leaves are pubescent and noticeably soft to the touch beneath, thus the species name *mollis,* meaning soft. Bright yellow flowers are borne in profusion from January to March when the twigs are bare. The leaves usually turn a golden yellow in autumn. This is the most commonly grown witch hazel in Vancouver, but deserves to be more widely planted, as it gives us the feeling that winter is nearing an end.

Probably the largest one in the city is a multi-trunked specimen on the north side of 70th Ave between Heather St and Logan St. There are two large ones growing together north of the Rose Garden in Stanley Park (just west of a tall *Hamamelis vernalis*), specimens on the SE corner of Cedar Cres and 19th Ave, one on the SE corner of 5th Ave and Dunbar St, and a large one on the NE corner of 9th Ave and Tolmie St; there are also several west of the Quarry Garden in Queen Elizabeth Park, in the Winter Garden at UBC Botanical Garden, and along the Rhododendron Walk at VanDusen Botanical Garden.

Hamamelis vernalis Sarg.
Spring Witch Hazel

This southeastern United States native is usually a shrub, rather than a tree. The leaves are narrower than other witch hazels, with only 3 or 4 pairs of veins, and they are usually glaucous and hairless on the undersurface. The yellow to red petals are less than 1 cm long, the shortest of any of the witch hazels. The usual forms seen in botanical gardens are low-growing, weeping plants with dull coppery red flowers, most commonly of the cultivar known as 'Lombart's Weeping.'

The typical wild form with yellow flowers is very rarely cultivated. There is an unusually large tree about 7 m tall in Stanley Park north of the Rose Garden, just east of two specimens of the slightly earlier flowering Chinese Witch Hazel, so that the two may be compared.

Hamamelis virginiana L.
American Witch Hazel

A deciduous tree from eastern North America growing to 5 m tall, the American Witch Hazel usually has multiple trunks. The leaves are nearly glabrous beneath and turn yellow in

September or October just as the pale yellow flowers are opening, thus masking the effect of the flowers. The bark has long been used medicinally as an astringent.

There are large specimens in the Old Arboretum on the UBC campus, one on the south wall of the Faculty Club at UBC, two in the Physick Garden in UBC Botanical Garden, one NW of the Stanley Park Dining Pavilion, and several in VanDusen Botanical Garden along the Rhododendron Walk and in the Eastern North American Section.

Liquidambar formosana Hance
Formosan Sweet Gum

This is the Asian counterpart of the more common Sweet Gum. It is native to southern China and Taiwan (formerly Formosa, thus the species name), where it can become a large tree growing to 40 m tall. The attractive leaves are very regularly 3-lobed, quite different from the 5–7-lobed leaves of the American species. The trees often have good late autumn colour. Formosan Sweet Gum is a very attactive tree that should be more popular.

It is rarely cultivated here. The only ones seen are a young one on the south side of the pond in Dr. Sun Yat-Sen Park, a larger one in the Asian Garden at UBC Botanical Garden, one outside the Asian Garden along SW Marine Dr just north of 16th Ave, and one at the end of the Rhododendron Walk in VanDusen Botanical Garden.

Liquidambar styraciflua L.
Sweet Gum

This abundant tree of the southeastern United States is considered a weed tree because it tends to grow profusely in old fields and vacant lots. Outside its natural range, it is considered a most desirable deciduous tree. The leaves are maple-like in shape, but are borne alternately along the stem, rather than in opposite pairs as in the maples. The small seed capsules, which hang in rounded clusters on stems, are much like those of the sycamores or plane trees, but the seed capsules are rarely produced in our climate. It is one of the best colouring trees in the Pacific Northwest, often not turning until November or even later. There is a tendency for some individuals to have corky ridges on the limbs. Only in recent years has this tree become popular in Vancouver.

There are no very old specimens in the city, but there are many smaller trees. There are rows along the north side of 12th Ave from Yew St to Vine St, on 47th Ave between Ontario St and Quebec St, on the UBC campus on the west side of the old Health Sciences Bldg on East Mall (across from the bookstore), and in the parking lot for the Miniature Railways in Stanley Park (the southernmost tree in this row has extremely corky twigs). A unique individual on the UBC campus, between Woodward Library and one of the Medical Sciences Buildings, is usually evergreen or loses its leaves only just before new growth begins in the spring. It even holds its leaves during very cold winters, but they turn brown and look very unsightly. •47

'**Variegata**' – The leaves of this cultivar are irregularly flecked and blotched bright yellow, or sometimes with half of a leaf green and the other half nearly yellow. There is a group of three (and a green-leaved form) just west of the Centennial Museum in Vanier Park, and two in the Eastern North American Section at VanDusen Botanical Garden.

Parrotia persica (DC.) C.A. Mey.
Parrotia or Persian Ironwood

This shrub or small tree grows to 12 m tall in its native Iran and the Caucasus Mountains. Small petal-less flowers, consisting of a mass of red stamens surrounded by brown furry bracts are produced on the bare twigs in late winter to very early spring. The trees are rather nondescript during the summer but are spectacular in the autumn when the leaves turn a variety of colours, from yellow and orange to pink and red, before dropping. The smooth grey bark becomes mottled and is an attractive feature as the trees become older. It makes a good, small street tree or specimen garden tree, especially if the lower limbs are pruned, forcing it into a premature tree form with only one or a few trunks.

There is a nice row of specimens outside VanDusen Botanical Garden on the north side of 37th Ave between Granville St and Oak St, some small ones in containers on the north side of the Bloedel Conservatory at Queen Elizabeth Park, one in the Old Arboretum at UBC, and some in the Asian Garden in UBC Botanical Garden. The largest one in Vancouver is probably the specimen on the north side of the Graduate Student Centre at UBC.

Hippocastanaceae – Horse-Chestnut Family

Aesculus × carnea Hayne
(*Aesculus hippocastanum* L. × *Aesculus pavia* L.)
Red Horse-Chestnut

Although very similar to the Common Horse-Chestnut, this tree is smaller, has more tubular, bright carmine pink to red flowers, and has fruits that are less spiny. There is a brilliant wash of colour on the ground beneath the trees for a few days after the flowers fall. The tree is assumed to be a hybrid between the Common Horse-Chestnut and the Red Buckeye from the southeastern United States, although its place and date of origin are not known.

It is less commonly planted here than the white species, but there are a few trees scattered around the city. The best specimens include ones along Osler St between Laurier Ave and King Edward Ave, along Cypress St from Hosmer Ave to Cedar Cres, along Hosmer Ave from Cypress St to Matthews Ave, in Queen Elizabeth Park below 33rd Ave, and a very large individual on the east side of Marguerite St between 29th Ave and 32nd Ave.

'**Briotii**' – This cultivar is a slightly more compact tree than the Red Horse-Chestnut with darker red flowers, but unless seen side-by-side the two are difficult to distinguish. 'Briotii' tends to have its peak of flowers a week or so after the more common Red Horse-Chestnut. It is rarely seen here, but there is a nice specimen on the NW corner of 33rd Ave and Willow St, one on the east side of Granville St between 49th Ave and 52nd Ave, one among a row of *Aesculus × carnea* on 14th Ave between Trutch St and Blenheim St, and one in the Mediterranean Section at VanDusen Botanical Garden.

Aesculus hippocastanum L.
Common Horse-Chestnut

This large deciduous tree grows to 30 m or more and is a native of the Balkan Peninsula, but it is now widely planted in temperate parts of the world. The large palmately-compound leaves with 5–7 toothed leaflets are distinctive throughout the summer. Thick twigs, fat sticky buds, and opposite pairs of large leaf scars are prominent winter features. The trees are outstanding in mid-May with large upright panicles of white flowers. Close examination shows that the flattened flowers are very attractive, composed of frilly white petals with a central yellow blotch that fades to rosy red. The fruits develop in the autumn, are covered with blunt spines, and bear one to three large, shiny, dark brown seeds, known to kids almost everywhere the tree grows as 'conkers.' The fruits are poisonous and should not be confused with the edible chestnuts (*Castanea*).

Horse-Chestnut trees are cultivated frequently as street or park trees in Vancouver. There are many individuals and rows of trees in the city, including large, old specimens around The Crescent and lining Osler St from The Crescent to Laurier Ave, along 14th Ave from Waterloo St to Arbutus St (mixed with the Red Horse-Chestnut), along 14th Ave from Crown St to Courtenay St, along Courtenay St from 11th Ave to 14th Ave, and two very large specimens on the south side of 16th Ave at Burrard St. An interesting individual can be seen at Jericho Beach at the corner of 2nd Ave and NW Marine Dr, opposite the Vancouver Hostel. This tree leafs out a month to six weeks earlier than all other specimens in the city and is often in full leaf by the end of March.

'Baumannii' (**'Flore Pleno'**) – This cultivar has very double white flowers, forming rounded, fringed balls. It seems to be rare here, although it may be mistaken at a distance for the wild white form and may be more common than is apparent. There are individuals below 33rd Ave in Queen Elizabeth Park and on the east side of Cypress St between Nanton Ave and Pine Cres.

'Plantierensis' – This sterile pink-flowered form is a cross between the hybrid Red Horse-Chestnut and its parent the Common Horse-Chestnut. The flowers are similar to those of the Red Horse-Chestnut, but they are paler pink and less tubular. There are two individuals in Queen Elizabeth Park below 33rd Ave with various other horse-chestnuts, and two on the east side of Balaclava St between 35th Ave and 36th Ave.

Hydrangeaceae – Hydrangea Family

Hydrangea paniculata Siebold
Tree Hydrangea

A native of southeastern China and Japan, this is our only cultivated *Hydrangea* to become a tree form. Many others are familiar garden shrubs with large showy blue, pink, or white flowers in summer. This species, too, is often considered a large shrub, but it quickly becomes a small tree, especially if the lower limbs are removed. It has very large, dense, pyramid-shaped panicles of flowers that begin to change from green to white in late summer, later turning pink, and finally golden brown as they fade. They remain on the ends of the bare branches and are distinctive throughout the winter. The flowers have a pleasant sweet fragrance. The pale green oval or elliptic leaves are rough to the touch. There are forms with tiny fertile flowers in the middle of a ring of 4-petalled sterile flowers, but the only form commonly cultivated here is the cultivar 'Grandiflora,' which has all sterile flowers.

There are many very picturesque tree-form specimens in the city, including large ones on 22nd Ave at Balsam St, on the north side of 21st Ave at Dunbar St, on the north side of 16th Ave between Blenheim St and Trutch St, on the sw corner of Highbury St and 18th Ave, on the west side of Oak St between 69th and 70th Ave, and on the NE corner of 7th Ave and Larch St.

Juglandaceae – Walnut Family

Carya cordiformis (Wangenh.) C. Koch
Bitternut Hickory

The hickories and pecans are common eastern North American trees, rarely cultivated in the West. Bitternut Hickory has pinnately-compound leaves with 7–9 slender, lanceolate, rather sickle-shaped leaflets, superficially more like those of the Black Walnut (*Juglans nigra*), except there are fewer leaflets. Over-wintering buds are long and very prominently bright yellow. The bark is tight and does not peel away as it does in the Shagbark Hickory. Rounded nuts are thin-shelled, with the shell splitting only about half-way to the base. The nuts are bitter and are not usually eaten.

It is rare here. The only tree found in the city is a relatively nice specimen (old enough to bear nuts) near the entrance to Totem Park Residences at UBC.

Carya ovata (Mill.) C. Koch
Shagbark Hickory or Shellbark Hickory

This large, common tree of the eastern North American deciduous forests is only slightly more commonly cultivated in the West than the previous species. It is similar to the Bitternut Hickory except that it usually has only 5 broader leaflets. The thick-shelled, sweet nuts are not usually produced here but are often gathered and eaten in the East. Autumn colour is often a good golden yellow. The common name comes from the strips of bark that split away from the trunks of old trees.

The only trees that have been located in the city include two large ones planted in Stanley Park, one in a row of Common Horse-Chestnuts along the walkway SE of the tennis courts and another just SW of the Ceperley Picnic Area; one in Queen Elizabeth Park on the north slope above Midlothian Ave; a small one on the east side of Trimble St between 2nd Ave and 3rd Ave; and several near the BC Bldg in the Pacific National Exhibition grounds on Hastings St.

Juglans – Walnuts

There are about twenty members of this genus, producing some of the choicest of temperate woods and edible nuts. The core of the pith of the twigs has distinctive cross-partitions, which can be readily seen when the twigs are split lengthwise. The walnuts have large pinnately-compound leaves that are aromatic and have a clammy feeling. The nuts have a hard shell surrounded by a thick husk.

Juglans ailanthifolia Carr. (*Juglans sieboldiana* Maxim.)
Japanese Walnut

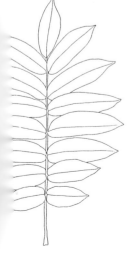

This very graceful, broad, flat-topped tree, with large pinnately-compound leaves, gives an almost tropical impression to northern landscapes. A native of Japan, it is sometimes cultivated here for its nuts or merely as an ornamental. It has 13–17 bright to pale green leaflets which can be up to 50 cm long, making them the largest leaves of any of the walnuts cultivated locally. The usually drooping leaves are borne on thick, stiff twigs. Each narrowly oval or lance-shaped leaflet is about 10–15 cm long and 5–6 cm wide, with minute teeth on

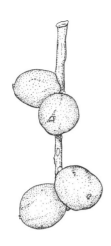

the edges, and soft fur and prominent veins beneath. The nuts are a distinctive feature, borne on long drooping stalks, often with 5 or more (sometimes up to 20) nuts, hanging chain-like on the tree. They are olive green in summer and turn brown in fall. There is a sticky (technically viscid-pubescent) texture to the nuts, giving them a distinctive fragrance when handled. Japanese Walnut is similar to, and often confused with, White Walnut, but the former has generally longer, more drooping leaves, averages more leaflets per leaf, and has much longer chains of nuts. The leaflets of Japanese Walnut appear entire unless the margins are examined closely for the tiny teeth.

It is probably the most commonly cultivated walnut in the city, but many of the young specimens are difficult to distinguish from young White Walnuts. There are a number of relatively large specimens, including two in front of the Agriculture Canada Research Station on NW Marine Dr at UBC, one on the NE corner of 41st Ave and Mackenzie St, one on the north side of 5th Ave between Alma St and Dunbar St (growing near an English Walnut), one on the north side of 3rd Ave between Collingwood St and Dunbar St, one in the Ceperley Picnic Area of Stanley Park, and several trees above the road on the NE side of Queen Elizabeth Park.

Juglans cinerea L.
Butternut or White Walnut

The Butternut and Black Walnut are the only two species of eastern North American walnuts. The Butternut gets its name from the sweet flavour of the nuts. The shell surrounding the nut is egg-shaped and the surface is covered with sticky hairs, which give the nut a distinctive fragrance when rubbed. The fruits begin to ripen and drop to the ground in September. The leaflets, usually 11–17 of them, are broader and less sharply pointed than those of the Black Walnut. They are bright olive green in summer, turning golden yellow before dropping in good years, or dull brown and dropping prematurely in dry years. Vigorous trees are easily confused with the Japanese Walnut, but Butternut usually has fewer leaflets, with more prominent marginal teeth, and the nuts are produced singly or in groups of two or three.

There are two trees on the western corner of Jervis St and Nelson St (in the lane on Jervis St between Comox St and Nelson St), two on the east side of Blanca St between 11th Ave and 12th Ave, an old specimen on the north side of 42nd Ave between Blenheim St and Collingwood St, and a long street planting along Yew St between 37th Ave and 40th Ave.

Juglans nigra L.
Black Walnut

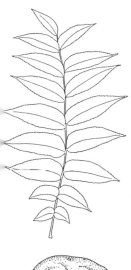

This is one of the most prized of eastern North American trees, valued for its wood used in making furniture and for its distinctively tasting nuts. The trees grow to 35 m tall and were common in fields and open forests in the East, but they are now becoming rare due to harvesting for lumber. The leaves are thinner and narrower than those of the other species cultivated locally. There are usually 15–19 leaflets with long narrow tips, with the terminal leaflet often smaller than the others. The trees are very late to leaf-out in the spring, often just showing green by early May. The fruits are usually produced singly or in pairs, are almost round (to about 5 cm in diameter) and are not sticky like the other walnuts. The shells are very hard and difficult to crack. Both the leaves and nuts have a pungent aroma when bruised.

It is fairly common here as an ornamental tree. There is a row of trees on the east side of East Blvd at Point Grey Secondary School, a large one west of the Rose Garden in Stanley Park, several trees on the western corner of Pendrell St and Broughton St, several on the western corner of Jervis St and Barclay St, a large one on the sw corner of Chancellor Blvd and Allison Rd, one behind the Physics Bldg at UBC, and one below 33rd Ave in Queen Elizabeth Park.

Juglans regia L.
English Walnut

This European walnut tree is the major source of the walnuts we eat. It is easily distinguished from the other species of walnuts by its fewer, broader leaflets (usually 5 but sometimes up to 9 leaflets). The smooth bark is pale grey. The smooth rounded or oblong nuts have been selected for thin shells, making them easier to crack.

There is a very large specimen on the NW corner of 5th Ave and Trutch St, one on the east side of Balaclava St between 2nd Ave and Point Grey Rd, one on the NE corner of Granville St and 16th Ave, several trees on 44th Ave between Killarney St and Earles St, a large one on the eastern side of Sasamat St between 9th Ave and 10th Ave, and a large one in the middle of the walkway near the entrance to the Children's Zoo in Stanley Park.

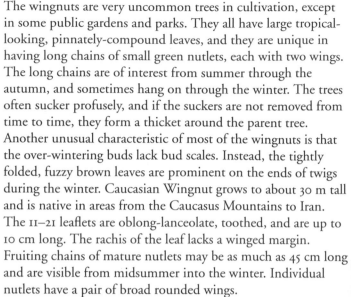

Pterocarya fraxinifolia (Lam.) Spach
Caucasian Wingnut

The wingnuts are very uncommon trees in cultivation, except in some public gardens and parks. They all have large tropical-looking, pinnately-compound leaves, and they are unique in having long chains of small green nutlets, each with two wings. The long chains are of interest from summer through the autumn, and sometimes hang on through the winter. The trees often sucker profusely, and if the suckers are not removed from time to time, they form a thicket around the parent tree. Another unusual characteristic of most of the wingnuts is that the over-wintering buds lack bud scales. Instead, the tightly folded, fuzzy brown leaves are prominent on the ends of twigs during the winter. Caucasian Wingnut grows to about 30 m tall and is native in areas from the Caucasus Mountains to Iran. The 11–21 leaflets are oblong-lanceolate, toothed, and are up to 10 cm long. The rachis of the leaf lacks a winged margin. Fruiting chains of mature nutlets may be as much as 45 cm long and are visible from midsummer into the winter. Individual nutlets have a pair of broad rounded wings.

There is a large individual on the western corner of Comox St and Chilco St, one in Maple Grove Park on the north side of sw Marine Dr at Maple St, one in the Old Arboretum at UBC, and two on the north side of the Education Bldg along University Blvd just west of Main Mall at UBC.

Pterocarya stenoptera C. DC.
Chinese Wingnut

This wingnut from China grows to 30 m tall and it suckers less than the other species. The rachis of the leaves is distinctly winged. The 11–23 oblong leaflets, to 8 cm long, are less prominently toothed than the Caucasian Wingnut. The racemes of fruits grow to about 30 cm long, and the wings of the nutlets are long, slender, pointed, and longer than the nutlet itself.

There is a row of trees on the east side of Place Vanier Residences and one in the Old Arboretum at UBC.

Magnoliaceae – Magnolia Family

Liriodendron chinensis (Hemsl.) Sarg.
Chinese Tulip Tree

This smaller Chinese relative of the North American Tulip Tree (*Liriodendron tulipifera*) is easily distinguished by its larger leaves with much deeper lobes, at least on the young trees seen locally. The flowers are smaller and less showy than the North American species, but are not produced on young trees. The tree grows to about 15 m tall in nature, much smaller than the North American species, although in cultivation it grows very quickly, and the young trees seen locally are already quite large.

It is rarely cultivated here. The only trees seen in the city are a young one in Dr. Sun Yat-Sen Park, a young but relatively large one in the Asian Garden of UBC Botanical Garden, and a number of young ones in the Sino-Himalayan Garden at VanDusen Botanical Garden.

Liriodendron tulipifera L.
Tulip Tree or Yellow Poplar

This elegant tree is the tallest of all trees in the eastern deciduous forest of North America, growing to nearly 70 m. It is an important timber tree in the East and an attractive ornamental tree, grown throughout the temperate regions of the world. The unusual leaf shape of this tulip tree is unlike that of any other tree. Tulip-shaped green and orange flowers, which give the tree its usual common name, are borne upright on the ends of twigs on mature specimens, in June in our area. They are not especially showy from a distance and may go unnoticed unless viewed from close range, especially from above (which is usually not possible on the tall trees). Seed clusters composed of winged fruits (samaras) are visible in winter on the bare branches.

There are a number of large specimens around the city. Probably the best known ones line both sides of 10th Ave from Alma St to Blenheim St, although these have been pruned and are not the typical shape. The finest old specimens are probably those on the north side of Harwood St between Bute St and Jervis St, on the NE corner of 50th Ave and Macdonald St, and on the SW corner of 29th Ave and Boundary Rd. There is a nice row of young trees showing the typical habit on the east side of Oak St between 59th Ave and Park Ave, two large specimens at the corner of 12th Ave and Ontario St, one at Yew St and SW

Marine Dr, and a street planting of relatively young (but typically shaped) trees on 5th Ave between Burrard St and Fir St. •48, 49

'Aureo-marginatum' – This rare cultivar has leaves with irregular yellow or yellow-green margins which are especially evident in spring, but they turn nearly uniformly green by midsummer. There are large specimens east of the Stanley Park Dining Pavilion, in the lawn west of Cecil Green Park at UBC, and a smaller one on the north side of the Pitch & Putt Golf Course in Stanley Park.

Magnolia – Magnolias

Magnolias have the largest flowers of our cultivated trees, and many of them are among the earliest of spring-flowering trees. Most *Magnolia* species and hybrids cultivated in our gardens are large shrubs or small trees, but some species have the potential of becoming very large trees at maturity. Most have large, thick, rounded or oblong leaves that drop in the autumn. There are several evergreen species, but they are generally rare this far north.

The over-wintering flower buds are covered with large, furry bud scales that slowly open revealing white or pink buds inside. The thick petals are technically known as tepals because magnolias do not have distinctive petals and sepals (petals and tepals are used interchangably in the following discussions). The tepals are often darker coloured on the outside and usually remain tightly closed on cold spring days, forming somewhat tulip-shaped flowers. On warm, sunny days they open out like flat saucers. Some have strong lemon-like or heavy fruity smells, especially on warm spring days or if brought indoors. The waxy stamens are often pink or red before shedding pollen. The fruits are elongated, knobby structures that split when ripe to reveal bright orange or red seeds. The seeds eventually dangle from the fruits on slender threads.

Many of the very large-flowered species do not flower well until they are some years old, which probably contributes to the fact that they are not grown more often. Also, some produce their flowers so early in the spring and have such large petals that they are often frosted or battered and broken by early spring rains and winds.

In addition to the species covered here, UBC Botanical Garden has a number of other Asian *Magnolia* species and some cultivars, many of which are very rarely seen in North American gardens. Recently, some of these have begun to flower for the first time.

Magnolia leaves

1 M. acuminata
2 M. dawsoniana
3 M. grandiflora
4 M. heptapeta
5 M. hypoleuca
6 M. kobus
7 M. quinquepeta
8 M. sargentiana
9 M. sieboldii
10 M. × soulangiana
11 M. sprengeri
12 M. stellata
13 M. tripetala
14 M. virginiana
15 M. wilsonii

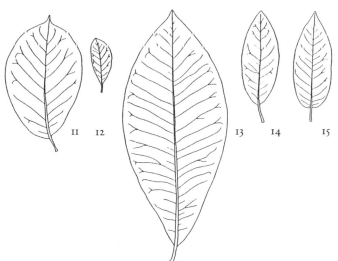

Magnolia acuminata L.
Cucumber Tree

This is a large deciduous tree, growing to 25 m tall, from the eastern deciduous forests of North America. Large oblong leaves are 10–20 cm long and half as wide. The flowers are among the dullest of an otherwise showy genus. Narrow yellow-green tepals are 5–8 cm long and may be described as interesting, at best. The common name refers to the elongated, knobby green fruits (7–10 cm long) which somewhat resemble small cucumbers. As with many magnolias, the fruits turn pink in autumn before exposing the orange-red seeds.

It is cultivated more as a novelty or in botanical collections than for any outstanding beauty. There is a large one in a backyard on the NW corner of 47th Ave and Macdonald St, and one in Stanley Park north of the Parks Board Office.

Magnolia dawsoniana Rehd. & E.H. Wils.
Dawson's Magnolia

A native of western China, this bushy deciduous tree, growing to 12–15 m tall, is the first of the large-flowered magnolias to flower in Vancouver. Buds first show as very dark clear pink, and the open flowers fade to pale pink, although there is variation in the colour of specimens in the city. The buds are distinctly long-pointed and there are fewer petals than in the slightly later larger-flowered species. The petals are slenderer and tend to curve to one side, especially before the flowers are fully open. Flowers usually peak at the end of March or, in cooler springs, the beginning of April. Specimens can put on an unusually good show of flowers, especially when previous summers are warm and sunny. Large, gnarled, dark pink fruits are often borne freely.

It is not a common tree. There are small specimens on the north side of 9th Ave between Blanca St and Tolmie St and on the north side of 46th Ave just west of Montgomery St (*Magnolia soulangiana* is also here and several *Magnolia kobus* nearby on Montgomery St for comparison); there are several around the Aquatic Centre at UBC, a large specimen on the SE corner of 41st Ave and SW Marine Dr (with a specimen of *Magnolia sprengeri,* which flowers later and has larger, darker pink flowers), a nice specimen in the Pine Woods at VanDusen Botanical Garden, and another nice one in UBC Botanical Garden's Asian Garden just north of the Moon Gate. The largest one in the city is certainly the one in Stanley Park, south of Lost Lagoon just east of the Pitch & Putt Golf Course. It is about 10 m tall and has slightly paler flowers than some of the others mentioned above. •50, 51, 52

Magnolia grandiflora L.
Evergreen Magnolia, Southern Magnolia, or Bull Bay

This is the great evergreen tree of the southern United States, where it grows to 30 m tall and bears hundreds of huge white flowers, smelling of sweet lemons, throughout the summer. The large evergreen leaves often have a coat of rusty hairs on the undersurface. The large, furry, cone-like seed-heads are attractive also. In our climate there is sometimes winter damage to the leaves. Flowers usually begin to open in mid-June and continue through the summer and early autumn in Vancouver. We are very near the northern limit of successful cultivation of the tree. The best ones here are those grown against a warm south- or west-facing wall, giving extra summer heat that encourages flowering.

There are a number of small trees around the city, including several relatively large ones in containers at the corner of Georgia St and Bute St, a recently planted row on Thurlow St at Melville St, several near the pool in Kitsilano Beach Park, several in and around the Dr. Sun Yat-Sen Classical Chinese Garden, and a number of relatively good individuals in the Pine Woods at VanDusen Botanical Garden. The most famous ones in Vancouver are the two in front of the Pacific Press Bldg on the east side of Granville St at 6th Ave.

Magnolia heptapeta (Buc'hoz) Dandy
(*Magnolia denudata* Desr.)
Yulan Magnolia

A deciduous magnolia from China reaching 15–20 m tall at maturity, the Yulan Magnolia is usually seen as a smaller tree or shrub in cultivation. Large oval leaves are about 15 cm long. The pure ivory white, slightly scented flowers are borne in profusion on bare branches in early spring, usually in late March in Vancouver. Petals are 12–15 cm long and are usually pointed upward, opening on warm, sunny days to a tulip shape. The Yulan Magnolia is one of the parents of the much more commonly cultivated Saucer Magnolia (*Magnolia soulangiana*) and can be distinguished by its pure white petals, usually lacking any flush of pink at the base. There are some pure white, or nearly pure white, forms of *Magnolia soulangiana,* and these are almost impossible to distinguish from *Magnolia heptapeta.*

It is fairly common here. There is one in Queen Elizabeth Park on the hillside above 33rd Ave near several *Magnolia kobus* (which flowers earlier and usually has fewer, much smaller flowers), two on either side of a *Magnolia soulangiana* on the north side of 33rd Ave between Carnarvon St and Blenheim St (where the two may be compared), one on the north side of 48th Ave between Oak St and Montgomery St, two nice ones on the north side of Point Grey Rd just east of Macdonald St, and one on the NW corner of 48th Ave and Cartier St.

Magnolia hypoleuca Siebold & Zucc.
Japanese White-Bark Magnolia

This large deciduous tree from Japan grows to about 30 m or more, with very large obovate leaves that are at least 30 cm long at full development. The species name comes from a combination of *hypo* (below) and *leuca* (white), referring to the colour of the underside of the leaves, which are both glaucous and pubescent. The large flowers (about 20 cm across) are produced after the leaves are fully developed, peaking in early May but with a few flowers later in summer. They open creamy white, fading quickly to brown-green on the inside and purplish on the outside, with red anthers. They have a strongly fruity, but unpleasant, smell that has been likened to a 'fruit salad in a barnyard.' Each flower lasts only a day or two.

There are a very few specimens in our area. There is one south of the monkey cages at the Zoo in Stanley Park (near the large Windmill Palm, *Trachycarpus fortunei*), individuals on the east and west ends of the Pitch & Putt Golf Course, a large specimen that flowers well every year against the west wall of the Graduate Student Centre along West Mall at NW Marine Dr on the UBC campus, four trees in a courtyard of the Heather Pavilion at Vancouver General Hospital just south of 10th Ave and east of Heather St, one in the Pine Woods at VanDusen Botanical Garden, and one in the Asian Garden of UBC Botanical Garden. •53

Magnolia kobus DC.
Kobus Magnolia or Great Magnolia

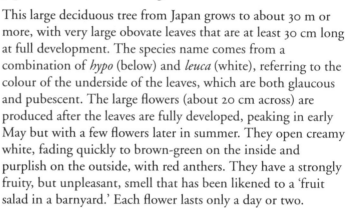

This tree, growing to about 25 m in its native Japan, has been planted only recently as a street tree in Vancouver. The trees have a very broad distinctive symmetrical outline. They generally do not flower well until they are older, although some of the plantings in Vancouver are now beginning to produce their white flowers, which appear just before the young leaves unfold in the spring. They are attractive, but certainly not among the showiest of magnolias. The first flowers usually open about the first week of March, peak by the third week, and are past by the end of the month. The trees are relatively non-descript during the summer, but an abundance of fruit may be produced in some hot summers (as in 1986 and 1987). These begin by looking like a small, twisted cucumber, turn bright rosy red by September, and later split to reveal brilliant orange seeds.

There are a number of relatively large trees in Stanley Park and Queen Elizabeth Park; there are street plantings along the south side of 12th Ave between Vine St and Arbutus St, along 7th Ave from Macdonald St to Balsam St, along 54th Ave between Oak St and Granville St, and along Maple St from 54th to 55th Ave; and there is a long row on Elliott St from 57th Ave to SE Marine Dr. •54, 55

Magnolia quinquepeta (Buc'hoz) Dandy
(*Magnolia liliiflora* Desr.)
Lily-Flowered Magnolia

This species is usually seen as a deciduous shrub, although it may become a tree growing to 4 m or more tall. The broad leaves are about 16 cm long. Flowers are borne before the leaves begin to grow, usually just coming into bloom in April as the *Magnolia soulangiana* cultivars are finishing. The form usually seen here is 'Nigra' with petals to about 10 cm long, dark purple on the outside and pale inside, forming long pointed buds opening to tulip-like shapes on warm days. There are often a few flowers produced later in the summer, but they are almost hidden among the large leaves. This species is one of the parents of the commonly cultivated *Magnolia soulangiana*.

It is not nearly as commonly grown here as the offspring *Magnolia soulangiana*, but there are a number of smaller ones around the city. There is a nice specimen with a tree-like form on the SE corner of 16th Ave and Vine St (usually at its peak of flower in early April), a nice small tree on the SE corner of 12th Ave and Stephens St, one on the north side of 14th Ave between Macdonald St and Stephens St, and some above the Rose Garden in Queen Elizabeth Park.

Magnolia sargentiana Rehd. & E.H. Wils.
Sargent's Magnolia

This deciduous magnolia from western China grows to about 20 m tall. It has large leaves (to 20 cm long), and its flowers are among the largest of all magnolias (up to 30 cm across). The petals are pale to deep pink on the outside and paler pink to white inside. The flowers are borne facing outward or nodding on the branches and are quite floppy when fully open. Trees in flower appear as if someone has attached large pieces of pink cloth on the branches when viewed from a distance. The large flowers suffer if, when fully open, there are strong winds or rains. The peak of flowering is usually toward the end of March. Unfortunately, very few, if any, flowers are produced until the trees are 10–15 years old.

It is rare here. There are a few young specimens, including one at the eastern end of the Pitch & Putt Golf Course in Stanley Park, several at VanDusen Botanical Garden (including a relatively large one which has begun to flower well), and several nice young, but relatively large, specimens that usually flower well in the Asian Garden of UBC Botanical Garden.

A similar species, *Magnolia campbellii,* also from the Himalayas, has the same large leaves and flowers. It differs in that the petals are a bit shorter, the flowers are not quite so floppy, and the inner petals tend to remain cupped around the stamens after the flowers open. In *Magnolia sargentiana,* all the petals lie flat, spreading away from the centre of the flower. *Magnolia campbellii* is not known in the city other than for a number of specimens in the Asian Garden at UBC Botanical Garden and in the Sino-Himalayan Garden at VanDusen Botanical Garden. •56

Magnolia sieboldii C. Koch
Siebold's Magnolia

This Japanese and Korean magnolia is more often seen as a multi-trunked shrub, but it may become a small tree to 10 m tall, especially if only one trunk is allowed to grow. It is very similar to *Magnolia wilsonii,* and the two are easily confused, but the flowers of *Magnolia sieboldii* are slightly drooping or held at right angles to the stems and the leaves are very sparsely hairy beneath. In *Magnolia wilsonii,* the leaves are longer, narrower, and silky-hairy beneath, and the flowers are very pendulous beneath the twigs. Siebold's Magnolia shows its egg-shaped buds when the leaves appear in the spring, but the flowers don't open until early May when the broad, ovate leaves are fully developed. The petals are pure white, with the outer petals reflexed and the inner remaining cup-shaped around the pale pink stamens. The flowers have a pleasant lemony-earthy smell. There may be some flowers produced through the summer. The flowers are followed by slender, club-shaped, pink-purple fruits that open in the autumn to reveal orange seeds.

Among specimens in Vancouver are a row of five definite tree forms (with one *Magnolia wilsonii*) as a street planting on the north side of 57th Ave from Beechwood St to Arbutus St, several on the north and west side of the Stanley Park Pitch & Putt Golf Course, several in the Pine Woods at VanDusen Botanical Garden, several in the Asian Garden at UBC Botanical Garden, and a nice tree form in a garden on the NE corner of 30th Ave and Maple Cres. •57

Magnolia × *soulangiana* Soul.-Bod.
(*Magnolia heptapeta* [Buc'hoz] Dandy) × *Magnolia quinquepeta* [Buc'hoz] Dandy)
Tulip Magnolia

This hybrid originated from a cross made in France and it is the most popular deciduous magnolia because it blooms prolifically, with large, showy flowers in early spring, even when very young. There are a number of cultivated forms, many of them named cultivars, and all have tulip-shaped blooms appearing before the leaves, with the petals in varying shades of pink on the outside and paler pink or white on the inside. The petals are usually darker pink toward the base on the outside. The near-white varieties closely resemble the *Magnolia heptapeta* parent, but there is almost always a pink flush toward the base of the petals in *Magnolia soulangiana*. Mature specimens may reach 15 m tall and may be even broader than tall. They are usually multi-trunked, but some trees are trained into single trunked specimens. The peak of flowering in our area is usually at the end of March.

It is a very common flowering shrub or tree here and the most noticeable magnolia in the city when in flower in March and April. The large specimens include those on the east side of Osler St near The Crescent, on the north side of Cornwall Ave between Yew St and Vine St (in Kitsilano Park), on the north side of 33rd Ave between Carnarvon St and Blenheim St (growing between two specimens of *Magnolia heptapeta*), on 37th Ave between Cartier St and Granville St, and a very large one on the west side of Angus Dr just south of 32nd Ave. There

are near-white forms on The Crescent between Angus Dr and
Hudson St, on the north side of 48th Ave between Granville St
and Cartier St, and on the south side of 36th Ave between
Dunbar St and Collingwood St.

Magnolia sprengeri Pamp.
Sprenger's Magnolia

This deciduous magnolia grows to 20 m tall in western China.
The leaves are obovate and reach 16–20 cm long. Trees are
usually broadly oval in outline and do not usually produce
many flowers until they are several years old. The large flowers
(to 20 cm across) are produced on the ends of branches in late
winter to early spring, usually in March here. The petals are
bright pink on the outside and paler inside. They are very
showy but not quite as large, nor do they open as wide, as those
of *Magnolia sargentiana*. Before the flowers open, large furry
flower buds are visible during the winter months. The very
large pink-purple fruits are often produced in large quantities
and are very showy in the late summer and autumn.

It is very rare here outside of botanical gardens. There is a
specimen about 7–8 m tall on the SE corner of SW Marine Dr
and 41st Ave that is very noticeable when in flower, a smaller
one against a house on the west side of Connaught Dr at 34th
Ave, two specimens at VanDusen Botanical Garden in the Pine
Woods along with several in the Sino-Himalayan Garden, a
number of rather large ones in the Asian Garden of UBC
Botanical Garden, and several around the Pitch & Putt Golf
Course in Stanley Park.

Magnolia stellata (Siebold & Zucc.) Maxim.
(*Magnolia kobus* DC. var. *stellata* [Siebold & Zucc.] Blackb.)
Star Magnolia

The popular Star Magnolia is a deciduous shrub or multi-
trunked tree from central Japan, growing to about 8 m tall.
Many authorities on the genus now consider it to be a small,
narrow-petalled variety of *Magnolia kobus*. The habit of the
plant is rounded and compact, becoming a white ball of flowers
when at its peak of flowering in late March or early April. The
oblong leaves (12 cm long) are narrower than any of the other
deciduous magnolias growing here. Young specimens flower
freely, with masses of white flowers (6–8 cm across) opening
much wider than other magnolia flowers. There are 12–18 long,
slender petals, giving the starry effect.

It is very common locally, but is most often seen as a shrub. However, there are definitely some small tree-like forms, including two relatively large specimens in front of the Stanley Park Dining Pavilion and several in the Quarry Garden at Queen Elizabeth Park, in VanDusen Botanical Garden, in the Asian Garden of UBC Botanical Garden, and on the north side of 15th Ave between Trutch St and Discovery St.

Magnolia tripetala L.
Umbrella Magnolia

This eastern North American deciduous magnolia grows to about 15 m tall and is a native of the southern Appalachian Mountains extending from Pennsylvania to Alabama. It has the largest leaves (more than 50 cm long) of any magnolia in cultivation in Vancouver. Flowers are produced after the leaves mature, from about mid-May into June. They have slender, creamy white petals and a sweet lemon-like smell when first opening. The smell becomes unpleasant as they fade.

There is a tree in the lawn NE of International House on West Mall off NW Marine Dr at UBC, a relatively large one at VanDusen Botanical Garden in the Eastern North American Section, a small one on the western side of Queen Elizabeth Park along Cambie St, and one in a garden on the NW corner of 35th Ave and Blenheim St. The largest one in the city is probably the one hanging over the path on the eastern end of the Pitch & Putt Golf Course in Stanley Park, with a smaller one on the north side.

Magnolia virginiana L.
Sweet Bay Magnolia

This evergreen or late deciduous shrub or small tree grows to 20 m tall in swampy places in the southeastern United States, and is an attractive tree in climates hotter than ours. They survive here, but never look as good as they do where they receive long, hot, humid summers. The small (to 10 cm), thick leaves are very glaucous beneath, a distinguishing characteristic. White, fragrant flowers are about 8 cm across, but are not produced very often or in abundance in our climate.

It is rare here. There are young trees at the east and west ends of the Pitch & Putt Golf Course in Stanley Park, and one in the Eastern North American Section at VanDusen Botanical Garden.

Magnolia wilsonii (Finet & Gagnep.) Rehd.
Wilson's Magnolia

This Chinese Magnolia grows to about 10 m tall at maturity and is often confused with the similar *Magnolia sieboldii*. It was named for Ernest Wilson of the Arnold Arboretum, who was a prolific plant collector in China. The flowers open in early May, hanging straight down beneath the silky, pubescent leaves. The flowers have pure white petals, dark red-pink stamens, and a lemon-like fragrance. The petals are longer and narrower than those of Siebold's Magnolia, and they open very flatly. The small trees show to their best advantage when planted so that they may be viewed from beneath.

There are specimens at VanDusen Botanical Garden in the Pine Woods, on the east and west end of the Stanley Park Pitch & Putt Golf Course, in the Asian Garden at UBC Botanical Garden, and one in a street planting of *Magnolia sieboldii* on the north side of 57th Ave between Beechwood St and Arbutus St, where the two may be compared.

Another very similar species, *Magnolia sinensis,* is known in Vancouver only from a small tree in the Asian Garden of UBC Botanical Garden.

Moraceae – Mulberry Family

Broussonetia papyrifera (L.) Venten.
Paper Mulberry or Tapa-Cloth Tree

This member of the mulberrry family is interesting for several reasons, not the least of which is that it is tolerant of a wide range of climates, from tropical to relatively cold temperate. In the Orient and Polynesia, where the tree is native, the inner bark is used for making tapa cloth. The tree has large, dark green, rough-hairy leaves with quite variable shapes. They may be either oval or mitten-shaped or have two or more broad lobes, all on the same tree. Young vigorous trees seem to have the most variable leaves, while old trees produce mostly simple, oval leaves. Small round heads of female flowers and elongated catkins of male flowers are produced on mature trees. Fruits have not been seen in our climate.

It is rare locally. There is a good tree at the north end of the Asian Garden at UBC and three young trees in Dr. Sun Yat-Sen Park in Chinatown.

Broussonetia papyrifera leaves

Ficus carica L.
Common Fig

The edible fig, a native of the Mediterranean region, grows into a deciduous tree to 10 m tall. It is the hardiest member of a very large genus (about 2,000 species) of largely tropical and subtropical trees, shrubs, and vines. The thick, dark green leaves may become 20 cm long, with 3–5 prominent lobes. The smooth, pale grey bark and very thick twigs are distinctive in winter. The edible fruits do not always ripen this far north, but there are good crops in years when we have long, hot summers and autumns.

Edible figs are fairly commonly cultivated in our area, especially in the east end of the city where they are usually grown against a warm, south-facing wall. There are trees on the north side of King Edward Ave west of Inverness St, one on the NE corner of 8th Ave and Trafalgar St, and two good specimens on the north side of 8th Ave between Trafalgar St and Stephens St. Probably the tallest one in the city is a multi-trunked specimen trained against a wall in the lane on the east side of Yew St between York St and 1st Ave.

Maclura pomifera (Raf.) C.K. Schneid.
Osage Orange

This unusual tree is a native of Arkansas and Texas. It is widely naturalized in other parts of the United States and is cultivated elsewhere for its showy chartreuse-green, warty fruit that resemble a green orange in size and texture. The fruits exude a white latex when cut, typical of many members of the mulberry family. Oblong, pointed leaves are borne alternately along slender, thorny twigs. The flowers are insignificant yellow-green fuzzy balls produced in mid- to late June in Vancouver. Unfortunately, no fruits have been seen here. The tree does not grow well in our cool, wet climate and there is often tip die-back.

There is a tall, slender tree in the Old Arboretum at UBC and a young one in the Eastern North American Section at VanDusen Botanical Garden.

Morus alba L.
White Mulberry

Of several species of mulberries, this and the Black Mulberry are the only two grown locally, with the White Mulberry being the more common of the two. Assumed to be native to eastern Asia, White Mulberry has been cultivated for centuries. It is a very common street tree in the Mediterranean countries, where it is usually pollarded. The leaves are very glossy above and bright or pale green. They are usually ovate in outline and either simple, with teeth around the margins, or with one or two (rarely several) irregular lobes. Several different leaf forms can be found on one tree or on a single branch. Leaves may turn clear yellow in the autumn. The inconspicuous flowers are produced in drooping catkins, followed by white to pale pink or sometimes red fruits that ripen in mid- to late summer. These are edible and have been used in a number of ways, such as making jams, jellies, pies, or liquors. The leaves are the main food source for the commercial silkworm.

It is not very common locally, but there are a number of mostly small specimens scattered around the city. There are two (with a Black Mulberry) at the western end of the Rhododendron Walk in VanDusen Botanical Garden, individuals as a street planting on the south side of 27th Ave between Ash St and Heather St (with cherries and plums), and one on the west side of Wallace St between 33rd Ave and 34th Ave.

'Macrophylla' – This is a very large-leaved form, with leaves twice the size of the typical wild form. The leaves are generally unlobed, except for those near the tips of vigorous growth. There are two rather large trees on the NE side of Queen Elizabeth Park and one in the Food Garden of UBC Botanical Garden.

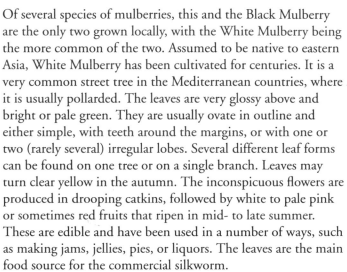

'Pendula,' Weeping White Mulberry – This is a weeping cultivar, often grafted on a standard trunk 2–3 m tall, thus allowing the grafted branches to weep gracefully. There is a specimen in the garden just north of the Stanley Park Dining Pavilion, one in the Children's Garden at VanDusen Botanical Garden, and one in the Food Garden at UBC Botanical Garden.

Morus nigra L.
Black Mulberry or Persian Mulberry

Very similar to White Mulberry, the Black Mulberry has darker green, usually larger leaves, with more prominent heart-shaped bases, and the leaves are more hairy beneath. The fruits vary from pink to dark red. It is a native of western Asia and is a smaller tree at maturity than the White Mulberry.

The only specimens found in the city are a fairly large one south of the Rose Garden west of Pipeline Rd (overhanging a walkway) in Stanley Park, and one with two White Mulberry trees near the western end of the Rhododendron Walk in VanDusen Botanical Garden.

Myrtaceae – Myrtle Family

Eucalyptus gunnii Hook.f.
Cider Gum

The familiar Australian genus *Eucalyptus* contains over 500 species of mostly grey-leaved evergreen trees and large shrubs. They are now a common part of the introduced flora and cultivated landscape in Mediterranean climates around the world. Unfortunately, Vancouver is just too far north for most of them to survive more than our mildest winters. Cider Gum from Tasmania is certainly one of the hardiest. Its flaking grey bark and drooping, hard, grey leaves are unlike those of any of our other cultivated trees. The small white flowers are petal-less, but they have a ring of relatively showy white stamens. Cider Gum flowers are borne in clusters of threes in late summer to autumn. Over the years of compilation of this book, many specimens of this species and some other relatively hardy gums (*Eucalyptus niphophila* and *Eucalyptus pauciflora*) have been observed in the city, but virtually all of these have died during our recent cold winters. However, small trees do grow so quickly that they may become tree-like in only a year or two after planting, so they have been included in this book. A few mild winters in a row will mean that they will certainly be replanted around the city.

There was a nice hardy individual in the Quarry Garden at Queen Elizabeth Park that has survived some cold winters (but was killed, at least to the ground, in the winter of 1990 – 1991), several smaller ones of this and other species in the Australasian Section of the UBC Botanical Garden's Alpine Garden and in the Southern Hemisphere Garden at VanDusen Botanical Garden.

Nyssaceae – Sour-Gum Family

Davidia involucrata Baill.
Dove Tree or Handkerchief Tree

A native of China, this tree attracts a lot of attention when in flower because of the large white bracts drooping from the limbs, hiding the small greenish white cluster of flowers. The leaves are heart-shaped, with large teeth around the margins, somewhat like some of the lindens. There are usually two bracts, one longer (about 15 cm) and one slightly shorter. The round flower cluster consists of a ball of greenish white stamens about 2 cm across surrounding a single pistil, and each develops into a long-stalked, brown, cherry-like fruit that hangs on the trees throughout the winter. Trees often do not begin flowering until they are well established and of some age.

It is not often cultivated here, but it has become more popular in recent years. The most famous Vancouver specimen is on the east side of SW Marine Dr just south of 49th Ave. When in full flower this tree looks as though there are a thousand handkerchiefs hanging from the limbs and covering the ground after they fall. There are two large specimens at Hycroft on the NW side of McRae Ave, two in the courtyard of the Biological Sciences Bldg at UBC, several north of the Agriculture Canada Research Station on NW Marine Dr near University Blvd, one in the western end of the Quarry Garden and one just north of the Quarry Garden in Queen Elizabeth Park, and a very tall specimen on the NE corner of the Pitch & Putt Golf Course in Stanley Park.

Nyssa sylvatica Marsh.
Black Gum, Sour Gum, or Tupelo

A large deciduous tree that has a wide range in eastern North America, Black Gum is one of the best trees for brilliant red autumn colour, although it is relatively nondescript the rest of the year. The tree has simple, shiny leaves and small green flowers, followed by pairs of small, black, cherry-like fruits, borne singly or often in pairs. The trees are tolerant of moist soils.

There are only a few small trees in Vancouver. There is one in the median of East Mall just north of University Blvd at UBC, and one on the south side of the Garden Pavilion of UBC Botanical Garden.

Oleaceae – Olive Family

Chionanthus virginica L.
Fringe-Tree

Usually seen as a large shrub or, rarely, as a small tree growing to about 10 m tall, this beautiful ornamental grows wild in the eastern United States from Pennsylvania to Florida and Texas. It is usually found in open forests or in the edges of forests. The nondescript oblong leaves sometimes turn golden yellow in the autumn. The small flowers are produced in May to early June and have 4 strap-shaped white petals that are borne in very large foamy clusters all around the twigs, just beneath the new, pale green leaves. The plants are very striking when in full flower. Flowers are followed by dark blue-black, olive-like fruits in autumn.

It is rarely grown here. There is one relatively large specimen near a large holly tree on the west side of The Crescent, one in a planter along Beach Ave at the foot of Davie St, one in front of the Barn Coffee Shop at UBC, and several young specimens in the Eastern North American Section at VanDusen Botanical Garden. •58

Fraxinus – Ashes

The sixty-five or so species of *Fraxinus,* commonly known as ashes, are a familiar part of the deciduous forests of the Northern Hemisphere. The combination of opposite, pinnately-compound (rarely simple) leaves and the distinctive, slender, one-winged fruits (samaras) make them easy to recognize. Most have very tiny flowers, but these are usually borne in large panicles, thus making at least some of the species moderately showy when in flower. They are valuable as timber and ornamental trees.

Fraxinus keys
1 *F. americana*
2 *F. angustifolia*
3 *F. excelsior*
4 *F. latifolia*
5 *F. ornus*
6 *F. oxycarpa*
7 *F. pennsylvanica*
8 *F. velutina*

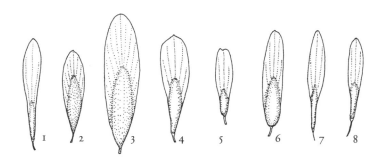

Fraxinus americana L.
White Ash

This beautiful ash grows to about 40 m tall and is a common component of the mixed deciduous forest of eastern North America, with a wide natural range from the Maritimes to southern Ontario and southward to Florida and Texas. There are usually 7–9 ovate or teardrop-shaped, pale green leaflets. Each leaflet has a distinct stalk, a characteristic that can be used to separate it from other ashes cultivated locally. The leaves are paler green beneath than above and the leaflets are usually toothless. The samaras, when produced, are narrower than those of other cultivated ashes. The autumn colour may be very showy, varying from yellow to red and wine, often all at once.

It is not a common ash locally. There is an old street planting of large specimens along 15th Ave from Oak St to Spruce St, and a row of younger ones along the north side of the Biological Sciences Bldg on University Blvd just east of Main Mall at UBC.

Fraxinus angustifolia Vahl
Narrow-Leaved Ash

Some authors consider this narrow-leaved European ash to include the very similar *Fraxinus oxycarpa*. Whether or not they are considered one or two species, there are definitely two different, rarely cultivated trees in Vancouver. This tree has the habit of Common Ash (*Fraxinus excelsior*), but has narrow

toothed leaves, usually with 9–13 leaflets that are glabrous above and below. The winter buds are dark brown. The trees tend to be broader and much 'twiggier' than *Fraxinus oxycarpa*.

Two large trees have been seen in the city: one along the road to the NE of the Stanley Park Zoo, and an old one in the edge of the Heather Garden at VanDusen Botanical Garden.

Fraxinus excelsior L.
Common Ash or European Ash

Our most common ash of parks and as a street tree, this large, valuable timber tree grows to 40 m tall and is a European native. It is without doubt the most variable of all ashes. There are a number of distinct forms and named cultivars, some of which are found locally. The sessile leaflets, usually from 7 to 13 in number, have very distinct teeth around the edges. The leaves are glabrous, except for small tufts of hairs along the base of the underside of the midrib of each leaflet. The keys are large and flat, with a wide wing extending past the body of the seed to the base. From late summer until spring the over-wintering, very dark brown or nearly black buds are another distinctive characteristic of this common ash.

There are many large trees in Vancouver. Three of the largest are at the entrance to Stanley Park (growing with a large *Liriodendron tulipifera*) between the causeway and the Rose Garden. There is a nice row of six large trees opposite the ball diamond near Ceperley Picnic Area in Stanley Park, several in the NW corner of Queen Elizabeth Park along 29th Ave at Cambie St, a planting on the north side of Devonshire Cres (opposite Devonshire Park) between Selkirk St and Hudson St, and a row on the south side of the Graduate Student Centre along Crescent Rd at West Mall on the UBC campus.

'Diversifolia,' One-Leaved Ash – A strange cultivar with variable leaves, but usually with only a single leaflet, and with large, irregular teeth. There are four trees with other ashes in the NW corner of Queen Elizabeth Park, a street planting of large trees along the south side of 41st Ave from Columbia St to Manitoba St, and a tree in the Fraxinus Collection west of the Heather Garden in VanDusen Botanical Garden.

'Heterophylla Pendula' – Combines the weeping habit of 'Pendula' with the typically simple leaves of 'Diversifolia.' There is a specimen along 33rd Ave near the European Beech Collection on the east side of Queen Elizabeth Park, and indivi-

duals at the entrance to and on the north side of the Pitch &
Putt Golf Course in Queen Elizabeth Park and in the Fraxinus
Collection in VanDusen Botanical Garden.

'Jaspidea' ('Aurea'), Golden-Twigged Ash – As the common
name indicates, the bright yellow twigs with contrasting dark
blackish brown buds are very distinctive in winter. It usually has
very good golden yellow colour in early autumn. It has been
planted as a street tree quite often in recent years, so there are a
number of young ones around the city. There is a street
planting on both sides of 59th Ave from Heather St to Laurel
St, several nice trees in Queen Elizabeth Park NW of the Rose
Garden and on the slopes north of the Quarry Garden, and
several young ones at the NW end of the Pitch & Putt Golf
Course in Stanley Park. •60

'Pendula,' Weeping European Ash – The weeping form with
long pendulous branches, with leaves typical of the wild form.
The tree is usually grafted on a straight trunk, thus allowing the
branches to droop to the ground. It is rare here. There is a nice
specimen in Stanley Park at the Zoo just SW of the duck pond
enclosure, one on the NE corner of 38th Ave and Yew St, and
one in the Fraxinus Collection at VanDusen Botanical Garden.

Fraxinus latifolia Benth.
Oregon Ash

The only Pacific Northwest native ash, the Oregon Ash is
common from California north to Washington. Records from
the west side of Vancouver Island are probably based on mis-
identifications of introduced *Fraxinus excelsior*. Oregon Ash is a
very prominent part of the deciduous tree vegetation in low,
relatively wet areas. The trees grow to about 20 m tall and are
either male or female. The females usually produce large
quantities of broad keys, with blunt or notched tips. The 5–7
sessile leaflets are usually broader than most of the other
cultivated ashes and are usually untoothed or obscurely
toothed. They are pale green above and even paler beneath.
They are very pubescent beneath and sometimes above also –
one of the best characteristics for distinguishing this ash from
others, locally.

It is not often cultivated in Vancouver. There is a nice row on
the west side of the Frederic Wood Theatre on Crescent Rd
between Main Mall and West Mall at UBC, and three large trees
on the western side of Queen Elizabeth Park between the
parking lot and the main quarry. •59

Fraxinus ornus L.
Flowering Ash or Manna Ash

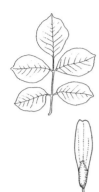

This is the most ornamental of the ashes with its large, dense panicles of fragrant creamy white flowers in May. Individual flowers are small, about 1 cm long, and are composed of 4 slender green-white petals. The 5–7 leaflets are usually rounder than those of most other ashes. The trees are only about 10 m tall at maturity. A native of southern Europe and Asia Minor, it was long a source of a sweet sap (manna), used as a laxative.

It is fairly common in the city. There are street plantings on the south side of 17th Ave from Trafalgar St to MacKenzie St (with some *Fraxinus excelsior*), on the east side of Willow St between 37th Ave and 39th Ave, on the north side of 31st Ave from Knight St to Inverness St, a group on the sw corner of the Agriculture Canada Research Station on NW Marine Dr at UBC, good specimens in The Crescent and in Kitsilano Park along Cornwall St west of Arbutus St, and a group (with a number of different ashes) on the sw corner of Queen Elizabeth Park along 29th Ave at Cambie St.

Fraxinus oxycarpa Willd.
Caucasian Ash

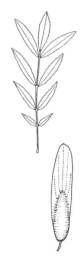

Native to the eastern Mediterranean, the Caucasus Mountains, and Asia Minor, this tree is very similar to the Narrow-Leaved Ash (*Fraxinus angustifolia*), and some authorities do consider them to be forms of the same species. The only form cultivated in the city is the following cultivar.

'Raywood,' Raywood Ash or Claret Ash – A cultivar that originated early this century in the Raywood Gardens, near Adelaide, Australia, this slender tree has narrow-toothed leaflets that give it a very feathery look in summer. The dark green leaves turn a deep reddish bronze in late summer and finally darker purple in autumn when they are most attractive, especially when used with other ashes that turn golden, such as *Fraxinus excelsior* 'Jaspidea.'

There are a few street plantings around the city, including a row (mixed with some broader-leaved *Fraxinus excelsior*) along Eddington Dr from Haggart St to Yew St (north of Prince of Wales Secondary School), a row along the north side of 29th Ave between Wallace St and Crown St (mixed with other ashes), and several young trees along 6th Ave and 7th Ave near Laurel St.

Fraxinus pennsylvanica Marsh.
Red Ash or Green Ash

The common names sound confusing, but both have been applied to different forms of this deciduous tree that grows to 25 m tall in nature in the eastern deciduous forests from southern Ontario to southern Saskatchewan and south to Florida and Texas. The relatively thick textured leaves usually have seven leaflets with small teeth around their edges. In cultivation in our area, the tree may be confused easily with the much more common European Ash (*Fraxinus excelsior*), but the Red Ash has a less prominent tuft of hairs along the midrib of the underside of the leaflets, the winter buds are pale brown (compared to the very dark brown or black buds of European Ash), and the keys are more slender and less prominently winged, more like those of the White Ash, *Fraxinus americana*. There is often good yellow autumn colour relatively early in the season.

There is a long street planting of Red Ash along the south side of 57th Ave from Cambie St to Laurel St and on the north side of 45th Ave between Macdonald St and MacKenzie St, an old specimen in the Old Arboretum at UBC, and two in the Fraxinus Collection at VanDusen Botanical Garden.

'Variegata' – A broad columnar form with very irregular white margins on the leaflets. It usually does not look very good, as the white margins sunburn easily and become brown and twisted. All of the local individuals of this cultivar seem to be grafted on Common Ash which often sends out shoots of its plain green leaves from below the graft. There may also be green-leaved branches or individual leaves among the variegated top part of the tree. There are a few specimens in the city, including street plantings (all mixed with European Ash) along 11th Ave between Clark Dr and Woodland Dr, for several blocks along 54th Ave from Inverness St to Prince Albert St, and along Waverly Ave between Kerr St and Doman St; there are specimens in the Fraxinus Collection at VanDusen Botanical Garden, and one in Kitsilano Beach Park near the foot of Yew St.

Fraxinus velutina Torr.
Arizona Ash or Modesto Ash

This open, sparsely branched ash is a native of the southwestern United States and northern Mexico. It usually has pubescent twigs, and the leaves have three or five densely pubescent leaflets with irregular teeth around the margins. The usual form

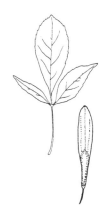

cultivated locally is a naturally occurring variety *glabra,* lacking hair on the twigs and leaves. It is generally unhealthy here and does not look good, although the sparse growth casts light shade and the leaves are small and do not present a litter problem when they drop – two characteristics that may be considered desirable.

The many plantings include rows on the north side of 12th Ave between Arbutus St and Yew St (alternating with *Fraxinus excelsior*), on 12th Ave between Trimble St and Discovery St, along Burrard St from Davie St to Barclay St and from Georgia St to Pender St, along Renfrew St from 1st Ave to Hastings St, and on 32nd Ave from Arbutus St to Haggart St.

Syringa vulgaris L.
Common Lilac

This eastern European native is a popular and commonly cultivated deciduous shrub or small tree to about 7 m tall, grown throughout temperate climates for its large clusters of strongly fragrant, white to purple flowers, produced in spring. The peak of flowering here is usually early May, but some individuals, especially the white varieties (which are also among the largest Lilac trees around), extend their flowering time for several weeks. The Common Lilac is not of top quality on our mild, wet coast, and flowering seems to be best after our coldest winters. There are scores of selected varieties, many of which may be seen at the Royal Botanical Gardens in Hamilton, Ontario, which has one of the largest lilac collections in the world. There are a number of other *Syringa* species that become tree forms, some of which are very hardy and are cultivated in cooler climates but not in our area.

It is usually thought of as a shrub, but there are a number of large, definitely tree-like specimens around the city, including a row on the UBC campus east of the tennis courts by Place Vanier Residences at NW Marine Dr and University Blvd, a large white, single-trunked specimen on the south side of 14th Ave between Columbia St and Manitoba St, a large white one on the east side of Alma St between 2nd Ave and 3rd Ave, a pale purple one on the north side of 53rd Ave between Ross Ave and Fraser St, and several white ones on Fraser St between 51st Ave and 53rd Ave.

Platanaceae – Plane Tree Family

Platanus × acerifolia (Ait.) Willd.
(*Platanus occidentalis* L. × *Platanus orientalis* L.)
London Plane Tree or Sycamore

This is a very popular street and park tree worldwide, but its exact origin is unknown. It obviously arose in cultivation, as the two parents are very widely separated in nature. *Platanus occidentalis* is native to eastern North America and has not been seen in Vancouver, but *Platanus orientalis,* which is from Eurasia, is grown locally, although it is not as common as the hybrid. The hybrid trees grow to 40 m tall and have bark that is beautifully mottled in shades of yellow, buff, and grey which is shed in irregular plates. Large, 3–5-lobed leaves are bright shiny green above and paler beneath. They are more shallowly lobed than those of *Platanus orientalis.* A unique feature among our local trees is that the base of the petiole is enlarged and surrounds the dormant bud of the next year. Round, soft-spiny fruiting clusters, 3–5 cm across, are borne on pendulous stems, usually with two per stem. The tree is tolerant of a wide range of city conditions and may be pruned severely without damaging the tree, but it does suffer from anthracnose, a fungal disease that causes leaves to become spotted and drop prematurely in summer.

It is very common here. The many plantings include those along MacKenzie St from Broadway to 16th Ave, along Broadway from Alma St to Wallace Cres, along 36th Ave from Carnarvon St to Blenheim St, on 42nd Ave from Main St to Ontario St, and on Laburnum St between Angus Dr and 63rd Ave; there is a row of large ones from the Zoo north to Lumberman's Arch in Stanley Park. •61

Platanus orientalis L.
Oriental Plane Tree or Sycamore

There is some doubt as to whether this tree is actually in Vancouver. There are a number of specimens planted with the hybrid London Plane Tree (*Platanus × acerifolia*) that appear to be Oriental Plane, although they may be merely hybrid seedlings that tend more toward that parent. The Oriental Plane has narrower, more deeply lobed leaves with much longer points to the lobes, and 2–7 fruits in a chain.

The only ones seen in the city are with plantings of the very common London Plane Tree, including a mixed group on the western side of Queen Elizabeth Park (west of the Rose Garden) and around the Ceperley Picnic Area in Stanley Park, three trees among many London Plane Trees in front of the University Hospital complex on the west side of Wesbrook Mall at UBC, and one of each kind on the west side of Lower Mall just north of Agronomy Rd on the UBC campus.

Rhamnaceae – Buckthorn Family

Rhamnus purshiana DC.
Cascara

This is a small tree found wild from British Columbia to California and Idaho, the bark of which has long been used medicinally as a laxative. One can still buy Cascara Sagrada in drugstores. The thick, elliptic leaves have about ten pairs of very prominent parallel veins. Vigorous shoots and young trees sometimes remain evergreen or partially so. The winter buds are distinctive because they lack bud scales, so that the bare tiny folded leaves are visible throughout the winter. Small greenish white flowers in spring are followed by purplish black cherry-like fruits about 1 cm in diameter.

Trees or small shrubs may be seen in the wild throughout the University Endowment Lands, Pacific Spirit Regional Park, and Stanley Park. It is rarely cultivated because it is not particularly ornamental, but is seen in parks as a small tree. There is a large street tree on the west side of Willow St between 14th Ave and 15th Ave (with hawthorns), a pair on the west side of Quarry Garden in Queen Elizabeth Park, several near the Forest Centre at VanDusen Botanical Garden, one just northwest of Woodward Library at UBC, and a very large multi-trunked specimen in the Old Arboretum at UBC.

Rosaceae – Rose Family

Amelanchier alnifolia Nutt.
Saskatoon-Berry or Western Service-Berry

This is a common shrub or small tree, especially in the dry interior and on rocky places near the coast in British Columbia. The leaves are nearly round with several large teeth toward the apex. Clusters of white flowers with 5 long, slender petals are borne in May, followed by blue-black fruits in summer. The fruits are extremely variable – some plants bear small, hard fruits, while others have very large, delicious edible ones.

It is very rare in Vancouver proper, although there are a few individuals in Pacific Spirit Regional Park and the University Endowment Lands. There is a relatively large tree form at Spanish Banks on the sw corner of Trimble St and Belmont Ave, a tall one east of the Stanley Park Pitch & Putt Golf Course, several relatively large ones in the bc Native Garden at ubc Botanical Garden, and a number of shrubby specimens on the nw rim of the Quarry Garden in Queen Elizabeth Park.

Amelanchier laevis Wieg.
Service-Berry

This is a large shrub or small tree growing up to 10–15 m tall at maturity in eastern North America. The white, 5-petalled flowers are borne in drooping clusters just as the young purplish bronze foliage is opening. The flowers are usually out in late March to early April and resemble those of the crab-apples, but the petals are much narrower and strap-shaped. The dark reddish black fruits are edible at maturity. There is sometimes good autumn colour locally. There are a number of other service-berries in eastern North America, many of which are difficult to distinguish from one another. It would seem that most or all of the tree forms cultivated in our area are of this species, although some may prove to be the very similar *Amelanchier canadensis*. There are many other common names in the East, including Shadbush, Shadblow, and Sarvice Tree.

There are a few relatively large specimens around the city, especially in parks. There is one slender tree in the Old Arboretum at ubc; and there are several around the Pitch & Putt Golf Course in Stanley Park, at VanDusen Botanical Garden (along 37th Ave and the parking lot, along Oak St, and in the Rosaceae Bed), on the east side of sw Marine Dr between Holland St and Crown St, and on the south side of Barclay St between Cardero St and Bidwell St.

Cotoneaster

This is a large and very commonly cultivated genus of evergreen or deciduous shrubs and small trees. Small, white to pink, 5-petalled flowers are followed by orange, yellow, black, or, most commonly, red fruits. Cotoneasters are usually seen in our gardens as ground covers or shrubs. However, some species do become distinctly tree-like with age, and a few of these are cultivated locally, usually as multi-trunked trees.

Cotoneaster salicifolia Franch.
Willow-Leaved Cotoneaster

A native of China, this evergreen shrub becomes a multi-trunked tree. It may be distinguished from the other tree forms of *Cotoneaster* by its narrow, willow-like leaves (thus the name *salicifolia*), which have the veins distinctly sunken below the upper surface of the leaf and soft pubescence beneath. It has large drooping clusters of small white flowers in mid-June, followed by many small (5 mm) bright red fruits.

It is commonly planted here as a shrub, quickly becoming a small tree. There are tree forms on the SE corner of the Stanley Park Dining Pavilion, on the east side of the Parks Board Office on Beach Ave, on the NW corner of 24th Ave and Collingwood St, along 10th Ave between Arbutus St and Vine St, a large one on the east side of Dunbar St between 3rd Ave and 4th Ave, and a row along Lower Mall near Agronomy Rd on the UBC campus.

Cotoneaster × watereri Exell
Waterer Cotoneaster

This is the name for a group of hybrids between *Cotoneaster frigidus, Cotoneaster salicifolius,* and *Cotoneaster rugosus.* They become large shrubs or small trees with semi-evergreen, broad leaves that are dark green on the upper surface and pale beneath. They may be hairy beneath when young, but become hairless with age. The reddish veins are distinctive on the lower side of the leaves. The plants may be completely evergreen during mild winters here or may become partially deciduous during colder winters. Flat-topped, drooping clusters of small white flowers are produced in early to mid-June, followed by coral red fruits, about 1 cm across, produced in large clusters that give a good show in autumn and winter.

There is a large multi-trunked specimen on the south side of 41st Ave at sw Marine Dr, one on the north side of Lagoon Dr SE of the Stanley Park Pitch & Putt Golf Course, and several west of the pool by the Children's Zoo ticket office in Stanley Park; the largest tree forms are in a row on the west side of Tatlow Park, along Point Grey Rd west of Macdonald St.

Crataegus – Hawthorns

This a very large and complex group of shrubs and small trees, to which about 1,000 names have been given. Their taxonomy is still in need of much work. They are found throughout cool northern temperate parts of the world, but especially in eastern North America. The red or orange fruits are commonly called haws, especially in Europe. Only a few trees are widely cultivated and even fewer are seen in the Pacific Northwest. Many are quite hardy in colder parts of North America. A number of species have long, stiff thorns. The leaves may be rounded to deeply lobed. Leaves on young, vigorous growth are often more lobed and with a prominent pair of stipules at the base. The white flowers have a typical rose family look, but they are often foul-smelling. Bright orange to red, or rarely, black fruit are borne in large drooping clusters, hanging on the bare branches well into winter.

Crataegus leaves

1 *C. × grignonensis*
2 *C. × grignonensis* with stipules from vigorous growth
3 *C. laeviagata*
4 *C. laeviagata*
5 *C. × lavallei*
6 *C. mollis*
7 *C. monogyna*
8 *C. phaenopyrum*
9 *C. × prunifolia*

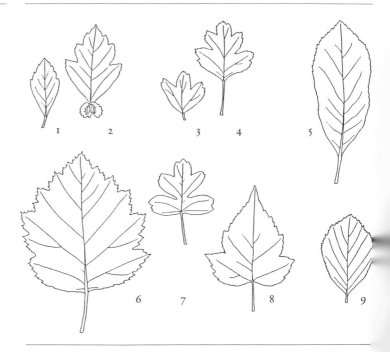

Crataegus × *grignonensis* Mouill.
(*Crataegus crus-galli* L. × *Crataegus pubescens* Steud.f.)
Grignon Hawthorn

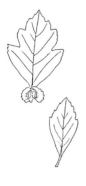

This nearly thornless, semi-evergreen hawthorn with an open, irregular growth habit was found in 1873 in a garden in Frankfurt, Germany. The leaves are variably shaped, ranging from nearly entirely to fairly deeply 3–5-lobed. The small orange-red fruits hang on for most of the autumn and winter months. It is an attractive small tree that deserves a more prominent place in our gardens.

The only specimens seen here are a large tree in front of Buchanan Bldg near the north end of Main Mall on the UBC campus and a smaller one in the Winter Garden of UBC Botanical Garden.

Crataegus laevigata (Poir.) DC.
(*Crataegus oxyacantha* L.)
English Hawthorn, Quick-Set Thorn, or White Thorn

There is a great deal of confusion and difference of opinion on the English Hawthorns – this species and *Crataegus monogyna* – especially the species to which the common cultivars belong. A majority of references seem to place most of them under *Crataegus laevigata*. The wild form, as found in Europe, has 3–5-lobed leaves (usually not lobed more than halfway to the midrib) that are glossy dark green above and duller beneath, with teeth at the lobe tips. Stipules, when present, are usually serrate. Also, the wild single (5-petalled) flowers of this species have 2–3 styles in each flower and, later, 2–3 seeds in each fruit, while those of *Crataegus monogyna* have a single style and single-seeded fruits. The two do hybridize when they grow together. Flowers are produced in late May to early June and have a foul smell. The wild form produces dark red fruit like small crab apples, in autumn. Double forms are sterile and do not produce fruit.

The wild, single form is fairly commonly planted here as a small street or park tree. Much more often cultivated is the bright rose red, double-flowered cultivar 'Paul's Scarlet.' A double white form ('Plena') and a paler pink form ('Flora Plena Rosea') are slightly less common. However, all of these forms are often inter-planted.

A group of the wild form may be seen at Spanish Banks on Trimble St at Belmont Ave, along Highbury St between 4th Ave and 5th Ave, and along Blenheim St between 6th Ave and 16th Ave (mixed with 'Paul's Scarlet' and 'Rosea Flore Pleno').

'Flora Plena Rosea' – A double-flowered form, paler than 'Paul's Scarlet' and slightly less commonly cultivated. In addition to the ones mentioned above, there are large ones on the north side of King Edward Ave from Hudson St to Oak St, one in front of the Faculty Club at UBC, and a long row on the west side of Western Parkway from Western Blvd to Chancellor Blvd (consists of a mixture of 'Paul's Scarlet,' 'Flora Plena Rosea,' and a few of the white wild form of *Crataegus monogyna*).

'Paul's Scarlet' ('Paulii,' 'Coccinea Plena,' 'Splendens') – This dark red-pink, very double-flowered form is commonly cultivated locally. In addition to the ones mentioned above, there is a large one in the median of King Edward Ave between Dunbar St and Collingwood St, and rows along 56th Ave from Fraser St to Prince Edward St, and 52nd Ave from Fraser St to Ross St (both rows alternating with plantings of Purple-Leaved Plum, *Prunus cerasifera*).

'Plena' – The double, white, sterile form, not often seen locally. There is a tree with a mixed row of 'Paul's Scarlet' and 'Flora Plena Rosea' on the south side of 13th Ave between Hemlock St and Alder St, one in the median of Angus Dr near The Crescent, and a row on 56th Ave from Prince Edward St to Ontario St (alternating with Purple-Leaved Plums).

'Rosea' – The single pink form. It is not as common here as are the white single forms and all the doubles. There are individuals on the SE corner of 3rd Ave and Yew St, a tall one on the NW corner of Angus Dr and Nanton Ave, and one on the west side of Stephens St between 3rd Ave and 4th Ave.

Crataegus × *lavallei* Herincq. ex Lav.
(*Crategus carrierei* Vauv.) (*Crataegus crus-galli* L. × *Crataegus stipulacea* Loud.)
Hybrid Hawthorn

This small deciduous tree of garden origin has a round-topped habit and dark green foliage. The trees are very late to leaf out in the spring and the white flowers are later than those of most other hawthorns, usually in late May to the first two weeks of June. Dark yellow to orange fruits hang in profusion on the trees into the winter. There are few thorns. The 5–10 cm leaves are oblong in outline, with small teeth on the margins, but lack the lobes found on many of our other hawthorns. The leaves are pubescent beneath and often colour yellow or orange in the autumn. The tree is more commonly grown to the south of us, where it produces huge crops of fruits.

There is a nice row of relatively large trees on the UBC campus at the north end of the Rose Garden at NW Marine Dr that are most noticeable in the fall and winter when the fruits are visible. There are two specimens (including a very large one) along King Edward Ave just east of Cypress St, seven trees on the north side of 13th Ave between Willow St and Heather St, and three along SW Marine Dr just east of Macdonald St. •62

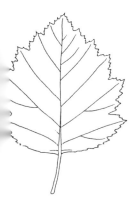

Crataegus mollis (Torr. & Gray) Scheele
Downy Hawthorn

A relatively large hawthorn (to 12 m) native from southern Ontario and Minnesota to Alabama and Arkansas, Downy Hawthorn is fairly commonly cultivated in parks and gardens for its large red fruits. The large, broad leaves, to about 10 cm long and often as wide, are among the largest leaves of any hawthorns. They are soft to the touch, especially beneath, and have large rounded teeth or lobes that are themselves again toothed. They are variable in size and shape on a tree or even on a single branch. White flowers in flat-topped clusters appear in May, earlier than most other hawthorns grown locally, are followed by large, vivid red fruits, 2–2.5 cm across.

It is very rare in cultivation here. There are specimens on the lane of the west side of Cartier St between Nanton St and 26th Ave, and one at the NW end of the parking lot for the Museum of Anthropology at UBC.

Crataegus monogyna Jacq.
Common Hawthorn, English Hawthorn, or May

This tree is very similar to *Crataegus laevigata* and some of the single forms listed under that species may actually belong here. The leaves are usually deeply 3–7-lobed, with the lobes reaching three-fourths of the way to the midrib. The lobes are usually less toothed than those of *Crataegus laevigata* and the stipules are often entire, or at least less serrate. The flowers have a single pistil and single-seeded fruits, as the species name *monogyna* indicates.

The only wild forms of this tree definitely found in the city are with a mixed planting of double forms of *Crataegus laevigata* along the west side of Western Parkway between University

Blvd and Chancellor Blvd, and a group on the sw corner of
Trimble St and Belmont Ave. There must be others around.

'Stricta' – The only cultivar grown in the city that seems to definitely belong under this species rather than *Crataegus laevigata*.
It is an open columnar form with sinuous branches, deeply
lobed leaves, and single white flowers fading to pale pink. It is
not common in the city, but there is a street planting on the
north side of 17th Ave from Laurel St to Willow St, one tree on
the NE corner of 5th Ave and Highbury St, two on the NW
corner of King Edward Ave and Alexandra St, and two on the
NW corner of 8th Ave and Granville St.

Crataegus phaenopyrum (L.f.) Medic.
Washington Thorn

In recent years, this hawthorn has become popular in central
North America as a small street tree. A native of the eastern
United States, it has glossy bright green leaves with irregular
lobes, the lowest usually being the largest. White flowers open
in mid-June in Vancouver. The fruits are a brighter orange-red
than those of many of our other hawthorns. The leaves sometimes have good autumn colour as well.

Although it seems to grow well here, Washington Thorn has
not yet caught on as a popular tree in Vancouver. The only
specimens seen include one on the west side of Fremlin St just
south of 48th Ave, and one each in the Eastern North American
Section and the Rosaceae Section at VanDusen Botanical
Garden. •63

Crataegus × *prunifolia* (Lam.) Pers.
(*Crataegus crus-galli* L. × *Crategus macracantha* Lodd.)
Broad-Leaved Cockspur Thorn or Plum-Leaf Hawthorn

This is assumed to be a hybrid between the two species given. It
is not known in nature, but is widely cultivated. There are a few
stout thorns, to 2 cm long, along the branches. The very shiny
dark green leaves are oval in outline with a row of sharp teeth,
and they often turn orange-red in autumn. The flowers are
produced in large fuzzy clusters and are followed by red fruit.

There are two large ones in the median of the parking area in
Kitsilano Park, one on the south side of 4th Ave west of
Highbury St (among a row of Weeping Willows), three on the
south side of Charles St at Kelowna St, and a grove on the SE
side of Queen Elizabeth Park. •64

× *Crataemespilus grandiflora* (Sm.) Bean
(*Crataegus laevigata* [Poir.] DC. × *Mespilus germanica* L.)
Hawthorn-Medlar Hybrid

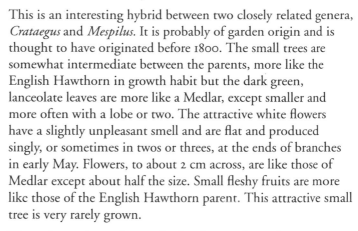

This is an interesting hybrid between two closely related genera, *Crataegus* and *Mespilus*. It is probably of garden origin and is thought to have originated before 1800. The small trees are somewhat intermediate between the parents, more like the English Hawthorn in growth habit but the dark green, lanceolate leaves are more like a Medlar, except smaller and more often with a lobe or two. The attractive white flowers have a slightly unpleasant smell and are flat and produced singly, or sometimes in twos or threes, at the ends of branches in early May. Flowers, to about 2 cm across, are like those of Medlar except about half the size. Small fleshy fruits are more like those of the English Hawthorn parent. This attractive small tree is very rarely grown.

The only specimens known in the city are three nice trees on the east side of Queen Elizabeth Park.

Cydonia oblonga Miller
Common Quince

This small deciduous tree grows to about 7 m tall and is a native of western Asia, although it has long been cultivated around the world in temperate climates for the large apple-shaped or pear-shaped hard yellow fruits used in making jellies and marmalades. The leaves are rounded or oblong and soft-furry. Pink buds open white, flushed with pink, to about 5 cm across, on the ends of branches in late April or early May (about the same time that apples are in bloom, which it closely resembles). The fruits are covered with a soft down and are fragrant when ripe, but unpalatable raw.

There is a nice small specimen on the north side of 11th Ave between Trimble St and Sasamat St, and large ones in the Rosaceae Section at VanDusen Botanical Garden and in the Food Garden at UBC Botanical Garden.

Eriobotrya japonica (Thunb.) Lindl.
Loquat

This evergreen tree is grown as an ornamental and as a minor economic fruit crop in subtropical climates. A native of the warmer parts of China and Japan, it grows to about 8 m tall, forming a round, bushy crown, with very distinctive, stiff

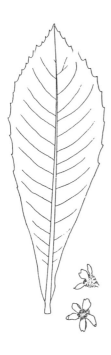

lanceolate leaves, 15–30 cm long, on short, thick petioles. The leaves are densely white-woolly beneath and glossy above, with large marginal teeth. Large panicles of small white flowers are produced in the autumn and winter, with fruits ripening in winter or spring in warm climates that are milder than Vancouver's. The fruits (3–4 cm long) are pear-shaped, the colour of apricots, and of a somewhat similar taste, with one or more large round seeds resembling a hazelnut. Occasionally, the fruits are seen for sale in local markets, especially in Chinatown. Vancouver is at the very northern limit for the trees to survive and there are a few small ones around. Trees from the coldest natural part of their range should be tried here. The local trees may flower in mild, long autumns, with the sweet fragrance often noticed more than the flowers themselves.

By far the largest one is in the north end of the Asian Garden of UBC Botanical Garden; there is a row of young trees in the lane SE of Denman St between Pendrell St and Comox St, and one at a home on the north side of Creelman Ave just east of Arbutus St.

Malus – Apples and Crab Apples

Malus includes the common edible apple (*Malus pumila*) with its large familiar fruits as well as a number of ornamental crab apples, with showy white to red-pink flowers and smaller edible fruits. Crab apples are much more common on the prairies, where they are hardy, than in Vancouver, where the ornamental cherries dominate. This is one of the most difficult groups of trees in which to sort out the ornamental cultivars. There are a few crab apples that are very distinct and easy to identify, but there are many more hybrids to which names cannot be easily given, especially a group of purple-leaved and flowered cultivars, here lumped under *Malus × purpurea*.

Locally, the crab apples are often assumed to be flowering cherries, but as a rule the crab apples flower a bit later than the cherries. They are usually small, round-topped or weeping trees with dull grey-brown bark and soft pubescent young twigs and leaves, the leaves having small blunt teeth. The bark and leaves of flowering cherries are both glossy, and there are very prominent sharp teeth around the leaf edges. The 5 rounded petals of the crab apples usually remain cup-shaped when fully open, compared to the flat and often notched or toothed petals of the cherries. The crab apples often produce an abundance of small, hard apples varying from green to yellow or dark red. None of the ornamental cherries produce fruits.

Crab apples suffer from a variety of leaf-spot diseases in our wet climate and the trees often defoliate by midsummer, this being one reason why they are not more popular here. Some individuals seem to be more resistant and the best ones are very showy when in flower and sometimes also when in fruit.

Malus leaves
1 *M. coronaria*
2 *M. floribunda*
3 *M. fusca*
4 *M. fusca*
5 *M. pumila*
6 *M. × purpurea*

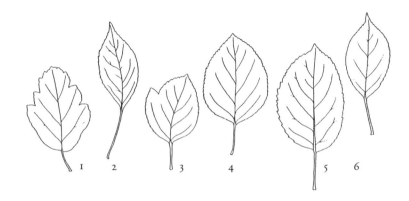

Malus coronaria (L.) Mill.
Sweet Crab Apple or Garland Tree

This is the latest flowering of all our local crab apples, usually in peak of flower from the middle to the end of May. The wild, single-flowered species is native to eastern North America and is not known in cultivation in Vancouver. The leaves are often 3-lobed or more, and the pale pink flowers have bright pink stamens. Fruits are green at maturity, 3–5 cm across, borne on long slender stems, and resemble large green cherries. •65

'Charlottae' – This semi-double cultivar is the only form of *Malus coronaria* cultivated in Vancouver. It has large drooping flowers of soft pink to nearly white with a blush of pink toward the base. The small trees are usually sparse in growth and leaf and are generally not very attractive, but individual flowers are beautiful. There is a group of trees on the west side of Queen Elizabeth Park below the Rose Garden and others on both sides of the eastern road up to the Quarry Garden, one on the west side of the tennis courts in Stanley Park, and a pair of trees on the east side of Blanca St between 2nd Ave and Drummond Dr.

Malus Floribunda Siebold ex Van Houtte
Japanese Crab Apple

Although generally treated as a species, the Japanese Crab Apple is thought to be an ancient hybrid. It is one of the most popular of crab apples because it consistently flowers profusely each spring. It is certainly the best crab apple locally. It is a dense, twiggy small tree with small narrow leaves. Flowers are on long pedicels, with showy round buds that are dark pink, opening to paler pink and fading to near white. The trees look best in bud or a combination of buds and early flowers. It looks a bit washed-out when all the flowers are fully open. The 2 cm wide crab apples are yellow when mature.

The many nice specimens in the city include those along Arbutus St from 57th Ave to sw Marine Dr, along 32nd Ave from Collingwood St to Dunbar St, along 2nd Ave from Alma St to Wallace St (alternating with Purple Crab Apple), on 19th Ave and 21st Ave from Trafalgar St to Arbutus St (alternating with Purple-Leaved Plum), and on both sides of 29th Ave from Heather St to Cambie St (also alternating with Purple-Leaved Plum). •66

Malus fusca (Raf.) C.K. Schneid. (*Pyrus fusca* Raf.)
Western Crab Apple

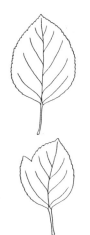

This small native tree is not known to be cultivated in the city, but it may be seen in the wild areas. It is a rather nondescript tree with either simple oval leaves or with one lobe near the base. The leaves sometimes give a good red colour in autumn. It has stiff twigs and clusters of small, white, fragrant flowers in April or May, followed by dark yellow to reddish brown fruits about 1 cm in diameter. These may hang on the bare branches into winter. It is often found growing in swampy ground. There has long been disagreement among botanists as to whether this tree should be considered an apple or a pear relative, as it is somewhat intermediate between the two.

There are several specimens on the sw corner of Trimble St and Belmont Ave at Spanish Banks, some in the edge of Pacific Spirit Regional Park along Crown St at the western end of King Edward Ave, and also some scattered in Stanley Park.

Malus pumila Mill.
Common Apple

The familiar apple is a small deciduous tree from southeast Europe and southwest Asia, long cultivated for its fruits. The leaves are round to oblong and soft pubescent, especially beneath. It is one of the last fruit trees to flower in our area. The buds are pink on the outside, opening to pure white, fragrant flowers. It usually flowers in mid-April, after the leaves are nearly fully developed.

Apples are very commonly cultivated throughout the city and are often found remaining in vacant lots and reseeded along roadsides. There are a number of trees in the park on the south side of Belmont St between Sasamat St and Trimble St at Spanish Banks, in the NW corner of Clarke Park (14th Ave and Woodland Dr), on the western side of MacMillan Bldg at UBC, in the Children's Garden at VanDusen Botanical Garden, and a fine collection of espaliered trees in the Food Garden at UBC Botanical Garden.

Malus × *purpurea* (Hort. Barbier) Rehd.
(*Malus* × *atrosanguineum* [F.L. Spath] C.K. Schneid. × *Malus pumila* Mill.)
Purple Crab Apple

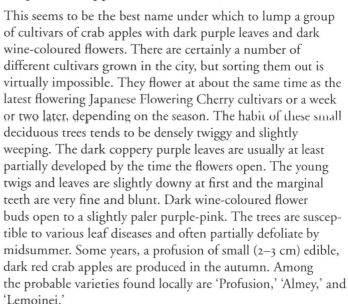
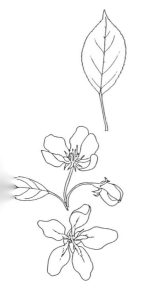

This seems to be the best name under which to lump a group of cultivars of crab apples with dark purple leaves and dark wine-coloured flowers. There are certainly a number of different cultivars grown in the city, but sorting them out is virtually impossible. They flower at about the same time as the latest flowering Japanese Flowering Cherry cultivars or a week or two later, depending on the season. The habit of these small deciduous trees tends to be densely twiggy and slightly weeping. The dark coppery purple leaves are usually at least partially developed by the time the flowers open. The young twigs and leaves are slightly downy at first and the marginal teeth are very fine and blunt. Dark wine-coloured flower buds open to a slightly paler purple-pink. The trees are susceptible to various leaf diseases and often partially defoliate by midsummer. Some years, a profusion of small (2–3 cm) edible, dark red crab apples are produced in the autumn. Among the probable varieties found locally are 'Profusion,' 'Almey,' and 'Lemoinei.'

The many specimens in the city include those along 2nd Ave between Alma St and Wallace St (alternating with *Malus floribunda*), on the west side of Angus Dr from 68th Ave to sw Marine Dr, on the south side of 4th Ave between Granville St and Hemlock St (under the south end of the Granville Bridge), on the NE corner of 4th Ave and Sasamat St, on the west side of Montgomery St between 48th Ave and 49th Ave, in the median of 16th Ave from Crown St to Blanca St, and on both sides of 21st Ave between Arbutus St and Puget Dr (alternating with Purple-Leaved Plum). •67

Mespilus germanica L.
Medlar

This rarely cultivated, small deciduous tree which grows to about 7 m tall is native to Europe and Asia Minor. The leaves are somewhat like an elongated apple leaf, to which it is related. The hard, olive brown fruits (to 5 cm across) are edible only after frost when they have begun to decompose. The flowers resemble those of the apple and open in late April. Medlar is one of the parents of the intergeneric hybrid *Crataemespilus,* discussed earlier in the book.

There is a large tree in the Old Arboretum at UBC, one espaliered against a south-facing wall of the squash court at Cecil Green Park off NW Marine Dr at UBC, and one in the Rosaceae Section at VanDusen Botanical Garden. •68, 69

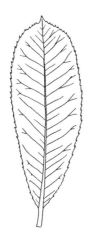

Photinia serrulata Lindl.
Sawtooth Photinia

This evergreen shrub or small tree from China grows to about 12 m tall. The thick, lanceolate leaves (10–18 cm long) have prominent marginal teeth. The new growth is bright copper or orange, becoming dark green later in the summer. Large, flat-topped clusters of small white flowers are borne in May and resemble those of the related *Sorbus* species. This species is not cultivated as often as the hybrid offspring *Photinia × fraseri,* which has been widely planted in recent years for its bright red new growth. *Photinia × fraseri* has smaller leaves with less prominent marginal teeth. Although the hybrid may ultimately become tree-like, most of those planted locally are still quite shrubby. There is one large individual of the hybrid on the north side of 2nd Ave between Arbutus St and Maple St.

It is rare here. There are two large specimens (with limbs to the ground) in the Asian Garden at UBC Botanical Garden, one

along Oak St near the Floral Hall at VanDusen Botanical
Garden, and a tall tree form against the south wall of the
Stanley Park Dining Pavilion.

Prunus – Cherries, Plums, and Other Fruits

In numbers of individual trees, diversity, and the spectacular
display of spring flowers, this is certainly the most important
group of trees in the city. The genus *Prunus* contains a variety
of different growth forms, from low shrubs to trees, from
edible fruit to fruitless (strictly ornamental) forms, and evergreen and deciduous forms. It includes economically important
fruit trees – cherries, plums, apricots, peaches, nectarines, and
almonds – as well as the ornamental flowering cherries and
plums and the evergreen cherries, the latter also known as
cherry-laurels or, locally, just as laurels.

Locally, there are both native and introduced members of the
genus *Prunus*. The flowering cherries and plums represent the
dominant part of our spring-flowering street and park trees.
Most of these do not produce fruit, although some fruit
production may occur in some individuals or in some years. In
general, the cherries may be separated from the plums by their
shiny bark with distinct horizontal rows of lenticels (corky
'breathing pores') on the trunk and major limbs. The plums
usually have duller bark, but there are exceptions to these
general characteristics. Of course, when fruits are produced, the
plums have relatively larger fruits on shorter stems than the
cherries. All wild species of *Prunus* have 5-petalled, white to
pink flowers, although many of the cultivated forms are double-
flowered and have many petals. The leaves of cherries are
generally simple, toothed, and glabrous.

Prunus 'Accolade'

A garden hybrid thought to be a cross between *Prunus sargentii*
and *Prunus subhirtella*, 'Accolade' is the most attractive of the
early-flowering ornamental cherries and deserves to be much
more widely planted in Vancouver. The buds are dark pink and
the semi-double flowers (with 10–12 petals) open to a shell
pink. Some trees may have a few flowers opening during warm
weather in winter, but most springs the peak of flowering is
about mid-March, 7–10 days later than the similar *Prunus
subhirtella* 'Whitcombei.' Some years, the flowers may peak as
late as the middle of April, almost at the same time as
'Whitcombei.' The tree seems to be slightly more resistant to
the fungal diseases that attack the leaves of other cherries.

It is not very common in Vancouver. There are several planted around Vancouver City Hall at 12th Ave and Cambie St, specimens in the Cherry Collection and along Oak St at VanDusen Botanical Garden, a group in front of the Old Barn Coffee Shop on Main Mall at UBC; there are street plantings along Yew St from 43rd Ave to 46th Ave, along 61st Ave from Ontario St to Manitoba St, and on Laurel St from 19th Ave to 20th Ave. It may be compared with the similar 'Whitcombei' on Marguerite St between King Edward Ave and 28th Ave ('Accolade' on the west side of the street and 'Whitcombei' on the east side), and on both sides of 50th Ave from Marine Cres to Macdonald St ('Whitcombei' is nearby, along the west side of Marine Cres and 49th Ave).

Prunus armeniaca L.
Apricot

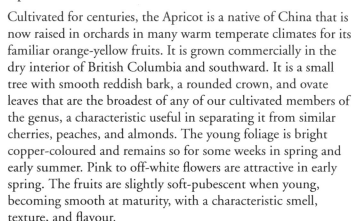

Cultivated for centuries, the Apricot is a native of China that is now raised in orchards in many warm temperate climates for its familiar orange-yellow fruits. It is grown commercially in the dry interior of British Columbia and southward. It is a small tree with smooth reddish bark, a rounded crown, and ovate leaves that are the broadest of any of our cultivated members of the genus, a characteristic useful in separating it from similar cherries, peaches, and almonds. The young foliage is bright copper-coloured and remains so for some weeks in spring and early summer. Pink to off-white flowers are attractive in early spring. The fruits are slightly soft-pubescent when young, becoming smooth at maturity, with a characteristic smell, texture, and flavour.

There are a few trees in back gardens around the city, but surprisingly few in prominent places. There is a tree against an apartment building on the NW corner of 4th Ave and Balsam St, and one in front of a house on the NE side of Davie St between Homer St and Richards St.

Prunus avium (L.) L.
Sweet Cherry

This commonly cultivated cherry is grown for its sweet fruits, but there are also selected forms that are grown for their ornamental value when in flower and, sometimes, for their autumn leaf colour. A native of Eurasia, it has very frequently escaped into fields and edges of woods on the lower mainland. The bark on old trees is smooth, reddish brown, and birch-like.

The mature leaves are dark green and relatively thin. Flowers hang on individual pedicels from the branches, and the peak of flowering is often around the second to third week of April.

Old specimens of the fruiting varieties around the city include a large tree on the east side of Cypress St between 10th and 11th Ave, one on the south side of Georgia St between Bute St and Jervis St, one on the south side of 54th Ave between Osler St and Montgomery St, and two on the east side of Hudson St between 49th Ave and 51st Ave.

'Plena' – This large tree is one of the most beautiful of all flowering cherries at its peak of bloom. It has a rounded or teardrop shape, and the limbs are covered with very double, pure white flowers drooping from every twig. The first flowers often appear around the 5 April and peak around the 20th. The leaves are narrower than those of the wild type of Sweet Cherry. It is sometimes planted as a street tree. There is a row of moderately large trees on the west side of Highbury St from 1st Ave to 4th Ave, on the north side of 29th Ave from Cambie St east to Dinmont Ave, and on 51st Ave from Fraser St to Ross St; there are some old specimens in front of the Canadian National Railways Station between Station St and Main St.

Prunus cerasifera J.F. Ehrh.
Cherry Plum or Myrobalan

The wild type of this plum has white flowers and green leaves but is not often cultivated here, unlike its purple-leaved, soft pink-flowered cultivar 'Atropurpurea.' It is native from central Asia to the Balkans. Five-petalled white flowers are borne in profusion before the leaves emerge, often in late February or early March in our area. The small (2–3 cm) red to yellow fruits are sweet and edible, although not often produced locally, probably because the very early flowers may not be pollinated or are frosted and killed. This wild form is often used as a rootstock on which the purple-leaved cultivars are grafted, and it is not unusual to see a vigorous sprout from the rootstock grow up among the purple-leaved graft, resulting in a mixture of white-flowered and pink-flowered, and later green-leaved and purple-leaved, limbs on the same tree. Obvious cases of sprouts from below the graft that have overtopped the graft may be seen on trees on the north side of 4th Ave between Highbury St and Wallace St, and on the SE corner of 3rd Ave and Blanca St. Many of the white forms in the city are probably a result of the pink-flowered top having died and the white understock sprouting from below the graft and then growing into a tree.

It is found infrequently here. There are white forms of Cherry Plum on the sw corner of 11th Ave and Macdonald St; several on the east side of Yew St between 38th Ave and 39th Ave, on the west side of Blanca St between 3rd Ave and 4th Ave, on the north side of Cornwall Ave just east of Cypress St, along 33rd Ave in Queen Elizabeth Park (these often fruit heavily); and a very large, slightly weeping form with several smaller, more erect, forms on the north side of 4th Ave between Trafalgar St and Blanca St.

'Atropurpurea' (*Prunus pissardii*), **Purple-Leaved Plum** – This is one of our most popular and widely planted small street trees. The pale pink buds open to even paler pink, almost white, 5-petalled flowers before the leaves emerge. The peak of flowering varies from the end of February to the end of March (or rarely early April), depending on the spring temperatures. Just as the flowers lose their petals, the leaves emerge, at first a light coppery colour but later turning a darker bronze purple for the summer and, finally, purplish red in the autumn. Very rarely, a good crop of dark red, edible plums is produced. There are probably thousands of this variety planted on our streets and in parks and private gardens. There are two trees in Kitsilano Park that begin flowering very early, usually a week or ten days before others around the city. The many large or especially showy plantings include those on both sides of 16th Ave from Granville St west to Arbutus St, along Cypress St from Broadway to Cornwall Ave, along both sides of 3rd Ave between Alma St and Wallace St, and along Blenheim St from 41st Ave to sw Marine Dr (with a few of the darker 'Nigra' mixed in). •70

'Blireiana' (*Prunus cerasifera* 'Atropurpurea' × *Prunus mumi*) – This cultivar is usually a small gnarled tree that never looks very healthy, at least in our climate. The flowers resemble those of 'Nigra' but are a week or two earlier, a slightly more intense pink, and double, with ten or more petals. The flowers are fragrant. It was more common locally, but has been replaced with other varieties in recent years. Plantings include those on the north side of 4th Ave between Trafalgar St and Balsam St, on both sides of King Edward Ave from Arbutus St to Macdonald St (alternating with the later-flowering *Prunus serrulata* 'Kanzan'), and alternating with 'Nigra' on 17th Ave from Laurel St to Oak St.

All three of these cultivars are especially abundant in the residential area bounded by King Edward Ave on the south, 16th Ave on the north, Arbutus St on the east, and Macdonald St on the west.

'Nigra' (including 'Thundercloud' and 'Vesuvius') – There are probably two or three cultivars of this darker selection in Vancouver, but they are virtually indistinguishable so are lumped here under 'Nigra.' It is not quite so common as 'Atropurpurea,' but is more attractive. It typically flowers about a week later than 'Atropurpurea,' with which it is often inter-planted. The buds and open flowers are a darker shade of pink and the leaves are almost black-purple. There are plantings along the west side of 4th Ave between Balaclava St and Bayswater St, on the south side of 12th Ave from Cypress St to Maple St, on 11th Ave between Blanca St and Sasamat St, and on 17th Ave from Laurel St to Oak St.

Prunus domestica L.
Common Plum

This familiar, edible plum is a small deciduous tree, probably from western Asia. However, it is now known only in culti-vation, where many varieties are grown for the popular fruits. Among the well-known cultivars are 'Damson' and 'Greengage,' as well as those used for making prunes. Older trees are often gnarled and of a picturesque form. The trees are ornamental in full flower when they are covered with small (less than 2 cm) creamy white flowers. Trees are usually in flower by the end of March or early April. When in leaf, the trees are very nondescript. The smooth-skinned fruit typically has a natural glaucous bloom on its surface.

It is very common here in gardens and sometimes escapes into wild areas. Among the older specimens in the city are two trees on the east side of Blanca St between 6th Ave and 7th Ave, one on the north side of King Edward Ave between Fraser St and Carolina St, one on the west side of Yew St between 20th Ave and 21st Ave, and three old specimens by the tennis courts north of the Aberthau Cultural Centre at 2nd Ave and Trimble St.

Prunus dulcis (Mill.) D.A. Webb
(*Prunus amygdalus* Batsch.)
Almond

A broad spreading tree growing to 10 m or more tall, with soft pink flowers, the Almond is native to western Asia and is widely cultivated for its edible nuts. The leaves are broadly lanceolate, and are very similar to those of the edible Peach. The fruit is different from all other cultivated members of the genus,

because the usually fleshy, down-coated outer layers are dry and inedible, splitting open at maturity. The kernel inside the stone is the edible almond of commerce, and, in some varieties, the source of almond oil. It is one of the earliest and one of the most attractive of the edible *Prunus* species, usually flowering the middle to third week of March, most years.

There are several trees on the NW corner of 33rd Ave and Cambie St, one on the NE corner of Elm St and 47th Ave, one on the SE corner of 53rd Ave and Granville St, one on the NE corner of 64th Ave and Fremlin St, and two large trees on the west side of MacMillan Bldg at UBC.

Prunus emarginata (Dougl.) Walp.
Bitter Cherry

This is a common deciduous tree native to the Pacific Coast from British Columbia to California. It goes unnoticed most of the year, with rather nondescript bark, oblong or broadly lanceolate leaves, and dull fruits. But, it is most noticeable when in flower, although it cannot begin to compare with the cultivated ornamental cherries. The rounded clusters of dull white flowers first begin to show colour during the middle to end of April and are fully out by the end of April to mid-May. As the common name indicates, the fruits are very bitter and are not eaten by humans. The tree is generally considered undesirable and is not often cultivated.

It is very common in the edges of native forests in our area and may be seen throughout Stanley Park, Pacific Spirit Regional Park, and the University Endowment Lands. The large specimens include one north of the Stanley Park tennis courts and one west of the parking lot for the Children's Zoo, and a large one on the south side of 4th Ave between Larch St and Stephens St.

Prunus × *hillieri* Hillier
(*Prunus incisa* Thunb. × *Prunus sargentii* Rehd.)
Hillier's Flowering Cherry

'Spire' – This is the only selection of this hybrid that seems to be grown locally. The trees form a stiff, narrow cone shape. The broad leaves feel soft to the touch, especially when young, and are especially hairy on the veins beneath. The leaf margins have long prominent teeth, each with a smaller tooth or two, and there is a prominent pair of glands at the base of the leaf blade. The pale pink flowers are partially double and have a prominent darker pink calyx. It flowers in late March to early April.

It is rare here. There are three nice trees in Memorial South Park on the south side of 41st Ave just east of Prince Albert St, and one in the NW corner of the Stanley Park Pitch & Putt Golf Course.

Prunus laurocerasus L.
Cherry Laurel or English Laurel

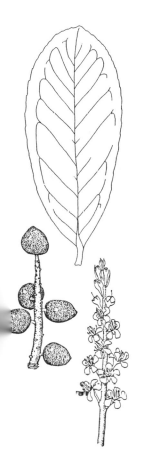

This common broad-leaved evergreen from southeastern Europe and southwestern Asia is probably the most over-used hedge plant in the Vancouver area. The typical form has pale green, very broad leaves and is used as a large hedge, usually just called laurel by most locals. Plants in our area do not often flower freely, but the long spikes of musky-sweet-smelling white flowers may attract attention in early April. If not pruned regularly, or not frozen back by our most severe winters, it quickly grows into a tree form with several quite large trunks, although it usually retains its lower limbs, so that the trunks are hidden. It is certainly borderline as a tree, it has been included in the list because it grows quickly. There are a number of commonly grown cultivars that have a low shrubby habit and narrower, darker green leaves than the wild form.

There is a multi-trunked tree-form on the north side of 13th Ave between Vine St and Balsam St, a large one that still has limbs to the ground on the SE corner of 4th Ave and Balaclava St, one on the NW corner of 5th Ave and Macdonald St, a large tree form on the NE side of Burnaby St between Thurlow St and Bute St, and several relatively large ones along 5th Ave between Larch St and Stephens St. A tall, single-trunked specimen on the NE side of Cedar Cres and 19th Ave is growing near a number of large rhododendrons, which it superficially resembles.

Prunus lusitanica L.
Portuguese Laurel or Portuguese Cherry Laurel

This species is not as commonly grown as Cherry Laurel, but it is a much nicer plant and should be more widely planted locally. It has smaller, darker green leaves with attractive red petioles and reddish young stems. It, too, is usually seen as a dense evergreen shrub or hedge, but specimens do eventually become small trees, especially if the lower limbs are removed.

Long, arching or drooping racemes of creamy white flowers with a musky-sweet fragrance are often produced in abundance in early to mid-June. These are followed by green fruits that become dark red-black at maturity, and astringent to the taste.

The specimens that are truly trees in the city include a nice 3-trunked tree in Tatlow Park along Point Grey Rd west of Macdonald St, one on the west side of Allison Rd between McMaster Rd and College Highroad, one on the west side of Highbury St at 6th Ave, and a double-trunked tree on the NE corner of Laurier Ave and Marguerite St. Certainly the largest trees in the city are three between the Zoo and the Dining Pavilion in Stanley Park.

Prunus mume Siebold & Zucc.
Japanese Apricot or Winter Plum

This is a small, winter- or early spring-flowering, deciduous tree that is widely cultivated in Japan. It is only just beginning to be known in North America, although there are a few older trees in the city. The young twigs are slender, but stiff, and bright green. The fragrant flowers, which usually flower in March here, are variable in size and colour, ranging from white to intense dark reddish pink, and are either 5-petalled or very double, with many petals. It is perhaps the most beautiful of flowering fruit trees, but is rather nondescript the rest of the year. Most of the ornamental varieties produce dry, inedible fruits, but there are a few fruit-bearing cultivars.

There are very few in the city. Probably the best individual is on the sw corner of King Edward Ave and Hudson St, which has relatively small, single flowers of an intense, dark red-pink. There are now small plants from this tree at VanDusen Botanical Garden. A slightly paler, but still shocking pink, cultivar with larger, double flowers that open very flat is on the west side of Fir St between 13th Ave and 14th Ave. There are two old specimens of a single white cultivar at Dr. Sun Yat-Sen Classical Chinese Garden, and several cultivars in the Winter Garden in UBC Botanical Garden.

Prunus 'Okame'
(*Prunus campanulata* Maxim. × *Prunus incisa* Thunb.)

This narrow, compact, small tree with stiff twigs produces flowers in great profusion along the branches, sometimes as early as late February. The semi-double flowers, produced in

threes, are a dark fuchsia pink at first, fading to mid-pink, and have deeply notched petals. The dark red-brown calyx and red stalks are noticeable. Fall colour is often good.

It is rare here. There is a relatively large specimen along the Rhododendron Walk in VanDusen Botanical Garden, and a row of small trees between the Asian Centre and the parkade on the UBC campus.

Prunus padus L.
European Bird Cherry

One of the latest flowering cherries, this European species is less attractive than many of the earlier flowering species and varieties, but is nonetheless showy in its own right. The shiny green leaves mature by early May, before the long penduous racemes of small white flowers appear in mid-May. The flowers are very similar to those of the common *Prunus serotina,* only twice the size.

It is rare here. There are three large specimens on the northeast slope of Queen Elizabeth Park between 33rd Ave and Midlothian Ave, and young trees along 16th Ave east of Wesbrook Mall. •71

Prunus persica (L.) Batsch.
Common Peach

Peaches, originally from China, are among the most popular small fruiting trees, being grown not only for their edible fruits but also for their attractive pink flowers. The long pointed leaves (to about 12 cm long) are the narrowest of all the cultivated *Prunus* species. The flowers vary from pale to dark pink and are usually produced in late March to early April. The fruits are usually covered with soft fur, except in the variety *nucipersica* (the Nectarine) which are smooth. Ornamental varieties have very double flowers, varying from white to dark pink, and usually do not produce fruits, or only small ones. These forms usually flower a few weeks later than the edible varieties.

The good ornamental double forms around the city include a dark pink on the NE side of Narvaez Dr between Puget Dr and 33rd Ave; a medium pink on the east side of Crown St between 16th Ave and 17th Ave and on the west side of Angus Dr at 59th Ave; a dark brilliant pink on the east side of Macdonald St between 22nd Ave and 23rd Ave and on the SE corner of 12th Ave and Trafalgar St; and a pale pink on the south side of 33rd Ave in Queen Elizabeth Park.

Prunus sargentii Rehd.
Sargent Cherry

This cherry is native to northern Japan, Korea, and Sakhalin and is often cultivated in Europe but is much rarer in North American gardens. It was named for Charles Sargent, the first director of the Arnold Arboretum in Boston. The typical wild form develops into a broadly triangular or round-topped tree growing to 25 m tall. Purplish pink, 5-petalled flowers in drooping clusters are produced in late March to April in our area, just as the dark coppery, prominently-toothed leaves emerge. Although the flowers are attractive, the best feature is the spectacular orange-red colour of the leaves early in the autumn. It is among the most reliable of local trees to put on a good autumn display.

The only wild forms in Vancouver seem to be a large specimen in Queen Elizabeth Park by the parking lot southwest of the Quarry Garden, and some young trees in the Asian Garden of UBC Botanical Garden.

'Rancho' – An erect form, not as tightly columnar as *Prunus serrulata* 'Amanagowa' but forming a narrow vase shape, often grafted at about 2 m high on a straight trunk. All the trees planted around the city are young and it will be interesting to see what they will look like in 50 years. Flowers are 5-petalled, mid-pink fading to very pale pink, and have a reddish pink calyx. There are usually flowers along the main limbs down to the point of graft. The young growth is coppery bronze, opening just as the flowers fade. Flowering is mid-season, usually between the early *Prunus subhirtella* cultivars and before most of the *Prunus serrulata* cultivars. Locally, street plantings include those along 57th Ave from Main St to Ontario St (with a few of a similar-shaped, white-flowered cultivar, probably a relatively new one called 'Pandora'), on 43rd Ave from Knight St to Inverness St, along the north side of 43rd Ave from Stirling St west toward Gladstone St, along Trutch St between 6th Ave and 7th Ave, and in Stanley Park east of the Pitch & Putt Golf Course.

Prunus serotina J.F. Ehrh.
Wild Black Cherry

This North American cherry is a deciduous tree growing to 25 m, native from Nova Scotia to North Dakota and south to Florida and Texas. Oblong leaves, 6–12 cm long, are glossy and relatively thick. Small white flowers are produced on long

catkin-like racemes, have a musky-sweet fragrance, and are often borne in profusion, even on young trees. These are followed by dark purple-black fruits, about 1 cm in diameter, which are edible when fully ripe.

It is fairly common locally as a street and park tree. There are street plantings along the north side of 63rd Ave from Fremlin St to Heather St, on both sides of 23rd Ave from Carolina St to Prince Edward St, on Cornish St between 68th Ave and 70th Ave, on the south side of 16th Ave from Camosun St for a half-block west, several at Locarno Park in the grove of trees on the NE corner of Belmont Ave and NW Marine Dr, and a group on the west side of the hill in Queen Elizabeth Park (with the similar, larger-flowered European Bird Cherry, *Prunus padus*).

Prunus serrula Franch.
Tibetan Cherry or Red-Bark Cherry

A relatively recent introduction into the Western horticultural community, this Chinese and Tibetan cherry was distributed by the Arnold Arboretum of Harvard University just after the turn of this century. It is still relatively rare in cultivation and deserves to be grown more widely. A deciduous tree reaching 10–15 m tall at maturity, its outstanding feature is the mahogany red bark that flakes and peels away from the trunk and main branches. Winter sun shining through the bare branches produces a wonderful effect. This is one of the best trees for people who enjoy colours and textures of bark. The small, pendulous creamy white flowers are produced around mid-April and are much less showy than are those of most other flowering cherries. They remain cup-shaped when fully open and have very long stamens. The dark green, lanceolate leaves are narrower and longer-pointed than most other ornamental cherries.

The known specimens in the city include one on the east side and two on the NW side of the Pitch & Putt Golf Course in Stanley Park, several along the Rhododendron Walk and in the Sino-Himalayan Garden at VanDusen Botanical Garden, two small ones on the SE corner of King Edward Ave and Valley Dr, and an individual on the SW side of Comox St between Denman St and Gilford St. •72, 73

Prunus serrulata Lindl.
Japanese Flowering Cherry

By far the most common flowering trees on Vancouver streets, and in parks and gardens, are the many forms of this species, or sometimes its hybrids with other cherries. The wild species is a

small tree growing to 20 m tall in China, Japan, and Korea, where it has been cultivated for centuries, but this wild form is rarely seen in North American gardens. The many cultivars make it one of the most variable and popular of our flowering trees. As a general rule, the flowers are pendulous, with 3–5 in a cluster, borne on a common, hairless stalk. There are often several small leaf-like bracts with fringed edges beneath the flower stalks. There are single forms with 5 petals, but most of the common cultivars have flowers with many petals, forming rounded pink or white 'pom-poms' when fully open. The leaves are pale green to coppery when young and usually only partially developed by flowering time. Mature leaves have distinct sharp-pointed teeth, sometimes elongated into fringes, and there is often a 'drip-tip' or long point on the end of the leaf. Fruits are very rarely produced. The trees are usually grafted on an erect trunk and often sprout from beneath the graft, so that some branches may bear different flowers. The habit of the trees may be very slender columnar or wide spreading to weeping, and everything in between. There are probably several dozen different cultivars grown in the city, but only about half-a-dozen are very common. A good project for someone would be to catalogue all of the cultivars in the city. The following are among the more popular and visible cultivars.

'Amanogawa' – A very strict, columnar form (like a small Lombardy Poplar). It is certainly one of the most distinctive cultivars of Japanese cherries. Older specimens, reaching 7–8 m tall, are slightly less strictly upright and often have a few sinuous arching branches. The young leaves are very dark green and appear with or just after the flowers open. The flowers are semi-double (with about 10 petals), dark pink in bud, opening to very pale pink, and slightly fragrant. The Japanese name means milky way. It is a fairly early cultivar, often in peak flower by late March or early April. There are many young specimens around the city, but not so many mature ones. The largest include a pair on the north side of Angus Dr between Hosmer Ave and Marpole St, a pair on the north side of 8th Ave between Fir St and Pine St, a row on the south side of 12th Ave between Oak St and Laurel St, and several on the east side of Vine St between 12th Ave and 13th Ave.

'Kanzan' ('Kwanzan,' 'Sekiyama') – Certainly the most popular of Japanese cherry cultivars in North America. Young trees have a very distinctive V-shape but become more spreading and more pleasing as they become older. The coppery leaves are at least partially developed by the time the flowers are fully out.

Dark pink buds open to very double pale pink 'pom-poms' and fade to mauve-pink. It is one of the later varieties to bloom and the flowers often do not last very long, but they are spectacular when in flower. The peak of flowers is usually in mid-April. This very common cultivar can be seen throughout the city. The showy plantings include those along 3rd Ave from Alma St to Wallace St (alternating with Purple-Leaved Plums), on 20th Ave from Trafalgar St to Arbutus St (also alternating with Purple-Leaved Plums), on 16th Ave from Blenheim St to Macdonald St (with the cultivar 'Ukon' in the median), along Georgia St from Stanley Park to Beatty St, on 33rd Ave from Macdonald St to Vine St, on Acadia Rd from Chancellor Blvd to University Blvd, and several old specimens in Nitobe Memorial Garden at UBC. •74

'Mikurama-gaeshi' – An early-flowering cultivar with a stiff V-shaped habit much like the later flowering 'Kanzan.' The leaves are coppery when young and the branches are covered with pale pink, single flowers. There is a notch at the tip of the petals. The flowers are usually fully out before the leaves, often by the end of March in many years, and are borne directly along the main branches. It is not very widely grown here, but there are three trees in Queen Elizabeth Park on the SE corner of Cambie St at 29th Ave, a row on 41st Ave between Granville St and Oak St (alternating with Purple-Leaved Plums), and along 18th Ave from Arbutus St to MacKenzie St (also alternating with Purple-Leaved Plums).

'Shimidsu Sakura' ('**Shogetsu**') – One of the late-flowering, more or less flat-topped, cultivars. This tree has green new growth, fringed leaves, and semi-double flowers in very long drooping clusters. Buds are pink, opening to pure white with 20–25 fringed petals, usually around mid- to late April. Young trees tend to be more upright and older ones more spreading. There are a number of specimens around the city, including street plantings on Trafalgar St between 34th Ave and 41st Ave (alternating with 'Kanzan'), several on a hill on the south side of the Student Union Bldg at UBC, several behind the Lutheran Campus Centre at Wesbrook Mall and University Blvd at UBC (with the similar cultivar 'Shirofugen'), and several on 35th Ave between Carnarvon St and Blenheim St (with 'Kanzan,' 'Shirofugen,' and 'Ukon' all within this block).

'Shirofugen' – Generally the last to flower and one of the most beautiful of Japanese cherry cultivars, 'Shirofugen' has a very flat-topped growth habit and very dark coppery young leaves that are usually partially developed by flowering time. The flowers are borne on long drooping stems. Buds are pink, opening to near white and fading to purplish pink, with 25-30 petals. It is very similar to 'Shimidsu Sakura,' but the darker coppery leaves and pinker flowers of 'Shirofugen' separate the two. It is usually at the peak of flowering at the end of April. Probably the nicest specimen is a large multi-trunked tree with broad spreading branches in a private garden at the corner of NW Marine Dr and Newton Wynd. There is a group of trees at Spanish Banks west of the snack bar, along NW Marine Dr in front of the Agriculture Canada Research Station at UBC, behind the Lutheran Campus Centre at Wesbrook Mall and University Blvd at UBC (with the similar cultivar 'Shimidsu Sakura'), and a grove in the median of King Edward Ave between Crown St and Wallace St.

'Shirotae' ('Kojima,' 'Mount Fuji') – A very showy cultivar with drooping clusters of semi-double, slightly fragrant, pure white flowers, it is a popular early to mid-season cherry. The habit of the tree is very flat-topped and spreading or drooping under the weight of the flowers, especially on older trees. The young leaves are pale green and have a distinct fringe around the edges. The peak of flowering is usually toward the end of March or early April. It is a common cultivar in the city. The nice specimens include those on the corner of 10th Ave and Alma St, a group just west of the Quarry Garden near the parking lot of Queen Elizabeth Park (with a group of *Prunus* × *yedoensis*), a large one in a garden on the north side of Chancellor Blvd between Acadia Rd and Allison Rd, and a large one on the west side of Western Forest Products Laboratory on NW Marine Dr at Main Mall at UBC.

'Tai Haku,' Great White Cherry – The habit of this tree is similar to, and it flowers about the same time as, 'Shirotae' but the spreading habit is not quite as flat-topped as 'Shirotae.' The young leaves are bright coppery-coloured and the flowers are large, single, and pure white. Mature leaves have long teeth on the margins. It is fairly common here. There are street plantings on 66th Ave from East Blvd to Adera St, on the east side of Blenheim St from 38th Ave to 41st Ave, on the north side of 39th Ave from Dunbar St to Blenheim St (with a row of *Prunus* × *yedoensis* on the south side of the street), and in the median of 16th Ave from Crown St to Dunbar St (alternating with 'Mikurama-gaeshi').

'Takasago' (*Prunus* × *sieboldii*) – There is some question as to whether this is a cultivar of *Prunus serrulata* or of hybrid origin. It is a relatively small, compact tree with V-shaped, gnarled limbs and coppery leaves. The leaves which are mostly produced after the flowers are fully out or just beginning to drop, so the trees usually appear leafless at the peak of flowering. Buds are dark pink, opening to pale pink, with semi-double flowers (12–15 petals). The flowers are not as large as most other *Prunus serrulata* cultivars and the flower pedicels are hairy. It does not do very well locally. There is a row on 38th Ave from Blenheim St to Dunbar St, on 60th Ave between Cambie St and Heather St (alternating with Purple-Leaved Plum), on 27th Ave from Crown St to Camosun St, and one tree on the east side of The Crescent.

'Ukon' – A very unusual and beautiful Japanese cherry with distinctively coloured flowers that are pale green in bud, opening to greenish white. Not showy from afar, the individual semi-double, drooping flowers are best seen at close range. The young foliage is coppery, usually partially developed by flowering time. It is a mid- to late season tree usually at its peak in mid- to late April. The unusual colour looks best alone rather than with the white or pink cultivars. There are a number of trees in Vancouver, including large ones in the triangle formed at the intersection of Macdonald St with 1st Ave and Point Grey Rd, a double row on Churchill St between 47th Ave and 49th Ave, specimens in the median of 16th Ave from Macdonald St to Blenheim St (with the cultivar 'Kanzan' on both sides of the street), and a pair of old ones in front of International House on West Mall at NW Marine Dr at UBC.

Prunus subhirtella Miq.
Rosebud Cherry or Higan Cherry

This bushy tree from Japan grows to about 8 m tall and is not known to occur in the wild although it has long been cultivated. It is thought by some experts to be of hybrid origin, having arisen centuries ago, either naturally or by an artificial cross. It is most often seen in cultivation as one of several distinct forms grafted on tall straight trunks. The buds open pale to dark pink and gradually fade to shell pink or white. There is a deep notch in the tip of the petals.

A few cultivars are commonly planted here in Vancouver. The first two attract a great deal of attention every year as they consistently begin to flower in the cool days of autumn. The unknowing observer assumes that they are spring-flowering cherries blooming out of season.

'Autumnalis,' Autumn Flowering Cherry – This is a relatively popular and common tree, although 'Autumnalis Rosea' is now more often planted. The trees are relatively flat-topped with arching branches. The flowers are soft pink in bud, opening to nearly pure white, with 10–15 petals, each with a notch at the tip. Flowering may begin during relatively mild weather at any time from late September or early October, just as the leaves are turning yellow, and continue throughout the winter into March. It does not produce the spectacular display that the spring-flowering cherries do, but it is nevertheless a welcome addition to our winter gardens. The largest trees in the city are several along West Mall at UBC, a group NE of the Quarry Garden in Queen Elizabeth Park, a row along 61st Ave from Manitoba St to Ontario St, and one on the NW corner of 13th Ave and Alder St. There were other plantings along 14th Ave between Sasamat St and Tolmie St and along 11th Ave from Macdonald St to Stephens St, but, unfortunately, the city decided to replace them with other species in 1989.

'Autumnalis Rosea' – This is slightly more common in cultivation than the previous cultivar. The growth habit is a bit more upright and the flowers open pink and remain so. At a distance these trees appear pink, while the cultivar 'Autumnalis' appears dingy white. There are trees in the median of Chancellor Blvd between Wesbrook Mall and NW Marine Dr, along the north side of King Edward Ave from Wallace St to Crown St, on the NE side of Pacific Blvd between Burrard St and Thurlow St, a group between the Parks Board Office and English Bay, and trees around the Pitch & Putt Golf Course in Stanley Park (with the darker pink, later-flowering cultivar 'Whitcombei,' which has 5 petals).

'Pendula,' Weeping Rosebud Cherry – This is a very popular cultivar, forming a graceful, weeping tree that resembles a small Weeping Willow, and which is covered in dark pink flowers in spring. It is usually grafted on an upright trunk. Old trees are very picturesque and unmistakable when in flower in March or April. The pendulous flowers are 5-petalled. There are large specimens on the SE corner of 15th Ave and Mackenzie St, on the west side of Granville St south of 28th Ave, on the north side of 33rd Ave east of Trafalgar St, on the north side of 45th Ave between Balaclava St and Carnarvon St, and on the east side of Adera St between 51st Ave and 52nd Ave.

'Pendula Plena Rosea,' Double Weeping Rosebud Cherry – Identical in form and general look to 'Pendula,' but the flowers are double, with more than 5 petals, and it usually flowers a week or so later than most of the single-flowered trees. There are trees on the north side of 33rd Ave east of Elm St, on the NW corner of Marguerite St and 49th Ave, and on the western edge of the Quarry Garden in Queen Elizabeth Park.

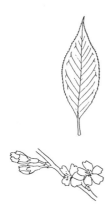

'Whitcombei' – One of the earliest of the spring-flowering cherries, the tree has a wide-spreading, flat-topped habit. The flowers are mid-pink, with 5 petals in drooping clusters. The trees have been severely damaged in recent years by a fungal leaf infection, which often causes all the leaves to drop after flowering. However, the trees seem to recover and continue to flower well each spring. There are four very nice individuals at the Agriculture Canada Research Station on NW Marine Dr at UBC, several large trees around the Pitch & Putt Golf Course at Stanley Park, a number of trees in front of apartments along Nelson St from Jervis St to Bute St, street plantings along 38th Ave from Ontario St to Quebec St and along 59th Ave from Main St to Prince Edward St, and three trees on the east side of Trafalgar St between 37th Ave and 38th Ave.

Prunus × *yedoensis* Matsum.
(*Prunus speciosa* Ingram × *Prunus subhirtella* Miq.)
Yoshino Cherry

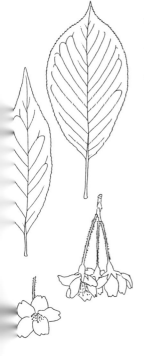

This is a very commonly planted old hybrid cherry, producing a profusion of flowers on wide-spreading, flat-topped trees. Flower buds are pale pink opening to nearly pure white or pale shell pink, with a slight sweet fragrance. There are usually 5 petals, although some flowers may have a few extra ones. The petals are strongly notched on their tips. The flowers are pendulous and usually borne 3–5 on a common, slightly hairy stalk. The flowers are often out in mid-March to mid-April, depending on the season, and do not last long, especially if there is windy or rainy weather.

This is the common blush pink, relatively flat-topped, mid-season cherry in Vancouver. There are hundreds or thousands of specimens planted here. There is a good group of four older trees at the corner of SW Marine Dr and Yew St, many in Stanley Park north of the tennis courts and around the Pitch & Putt Golf Course, and many in Queen Elizabeth Park around the parking area southwest of the Quarry Garden; there are good street plantings along the south side of 39th Ave between Dunbar St and Blenheim St (with a row of the *Prunus serrulata* 'Shirotae' on the north side), along Manitoba St from 39th Ave to 43rd Ave, and (at UBC) along Lower Mall and at the entrance to Nitobe Memorial Japanese Garden on NW Marine Dr. •75

Pyrus calleryana Decne.
Callery Pear

This ornamental tree has been very widely planted recently, especially in eastern North American cities and in California. A native of China, it is considered a good street tree because of its upright habit, broad shiny leaves, white flowers in early spring, good autumn colour of red and yellow, and resistance to the fire blight disease. The leaves are broader than those of the Common Pear and they may remain partially evergreen in warm climates. Wild forms have stiff thorny twigs, lacking in many of the cultivated forms, and the small fruits are only about 1 cm across. The winter twigs are a brassy colour. There are a number of popular cultivars. The first introduced was 'Bradford,' with a broad oval shape, but 'Chanticleer,' with a narrower habit, is more widely planted now.

There is a recent planting of small trees for a number of blocks along Great Northern Way between Main St and Clark Dr, and larger specimens of several cultivars are in the Asian Garden at UBC Botanical Garden and in the Rosaceae Collection at VanDusen Botanical Garden.

Pyrus communis L.
Common Pear

Common Pear is probably of hybrid origin and is an important commercial tree which has long been cultivated for its large, edible fruits. There are more than a thousand known cultivated varieties of pear! The leaves are the glossiest of all cultivated fruits, are oval or rounded, and often turn orange or red in the autumn. The thick twigs of most cultivated pears are usually thornless, but vigorous young shoots may have short, thick spur-like thorns. The trees are fairly ornamental in bloom, usually in late March to early April when they are covered with dull white flowers (compared with the purer white to pink-white of apples and cherries). The 5-petalled flowers are cup-shaped when fully open, have distinctive red anthers, and a musky smell. The tree is most easily identified in the summer and autumn when the large fruits are ripening.

It is commonly planted in gardens and orchards for its fruits, but not as an ornamental. Visible old specimens around the city include a large one in Vanier Park in the lawn just east of the Maritime Museum, one on the east side of Burrard St at the lane between 13th Ave and 14th Ave, one along the street with ornamental cherries on the south side of 60th Ave just west of Ontario St, one on the SE corner of 12th Ave and Oak St, one on the west side of Killarney St just north of 44th Ave, and one on the NW corner of 64th Ave and Buscombe St.

Pyrus elaeagrifolia Pall.
Olive-Leaved Pear

A very rarely cultivated, but attractive, small pear from Asia Minor. It has thorny twigs that are white and have a soft pubescence when young. The leaves are very grey-green above and soft, white beneath, and have long drooping petioles. White flowers are not especially showy against the grey leaves, but the overall effect of the tree is very pleasing. Small green pears, about 2 cm in diameter, are seen in the autumn. This tree deserves to be much more widely cultivated in our area.

The only specimen located is a relatively large one on the west side of the Mathematics Bldg between Agricultural Rd and Memorial Rd at UBC.

Pyrus salicifolia Pall.
Willow-Leaved Pear

This small deciduous tree from eastern Europe is cultivated for its graceful habit and ornamental, grey, willow-like leaves. The white flowers open by late March or early April. They are moderately attractive, although foul-smelling. Small, hard, green fruits, 2–3 cm in diameter, are produced in the autumn, but are inedible. The tree might easily be mistaken for the Russian Olive (*Elaeagnus angustifolia*) when not in flower or fruit.

Very rarely cultivated in Vancouver, this attractive tree deserves to be more commonly planted. The only specimens seen are of the weeping cultivar 'Pendula,' of which there are two nice trees on the east side at the bottom of the Quarry Garden in Queen Elizabeth Park, and two at VanDusen Botanical Garden (one in the Rosaceae Section and one at the western end of the Alpine Garden). •76, 77, 78

×*Sorbaronia alpina* (Willd.) Schneid.

(*Aronia arbutifolia* [L.] Pers. × *Sorbus aria* [L.] Crantz)

A garden hybrid between the shrub *Aronia* and the tree *Sorbus*. The hybrid itself is usually shrubby, except when grafted on a standard trunk (as are the only specimens in Vancouver). The puckered leaves most resemble a small *Sorbus aria* leaf, 10–12 cm long, with teeth all around the margins and a dense grey-green coat of hairs beneath. New leaves emerge very early in spring and are usually fully expanded by late March. White, flat-topped flower clusters are followed by dark dull red fruits, about 1 cm across, borne in drooping clusters on red stems partially covered with white pubescence.

This tree is not generally in cultivation other than in special collections in parks or botanical gardens. Probably the only trees in the city are three nice grafted specimens on the north rim of the main Quarry Garden in Queen Elizabeth Park, and one in the lawn just north of the Winter Garden in the UBC Botanical Garden.

Sorbus – Mountain-Ashes, Rowans, and Whitebeams

These members of the rose family are common in the Northern Hemisphere, especially in China and the Himalayas. Usually small trees or shrubs, they often have pinnately-compound leaves with toothed leaflets, large clusters of flowers that are strongly scented (pleasant to very foul), and attractive, often orange or red, fruit. The familiar European Mountain-Ash, so commonly grown in our gardens, has these characteristics. However, there are many members with simple leaves, such as the whitebeams, and many have fruits that are pure white, pink, yellow, and from pale orange to very dark red, and occasionally, brown.

There are a large number of other species in cultivation in the collection at VanDusen Botanical Garden and in UBC Botanical Garden, especially in the Asian Garden. Anyone interested in seeing the many other forms should visit these two collections.

Sorbus aria (L.) Crantz
European Whitebeam

This European native is so unlike many of our more familiar species, with its rounded, nearly entire leaves that have small, irregular teeth and a flocking of white down on the under-surface, that it is generally not recognized as a *Sorbus*. However, there are a number of other *Sorbus* species with similar simple leaves. The young leaves are white on both sides, but the upper surface soon becomes relatively smooth and hairless. The white flowers in flat-topped clusters in spring are followed by round, scarlet to brown fruit, about 1 cm in diameter. Some of the trees in the city are probably the cultivar 'Lutetiana,' in which the young growth is yellow-green, but there is so much variation in individuals that it is difficult to know definitely.

It is fairly common here as a street and park tree. There is a group of trees below the Rose Garden on the west side of Queen Elizabeth Park, a group of seven on the south side of the Queen Elizabeth Pitch & Putt Golf Course (by the parking lot), a row on the east side of Sidney St from 25th Ave to 27th Ave, a row on the north side of 54th Ave from Kerr St to McKinnon St, a row in the median of Cambie St south of 41st Ave, and a nice pair along West Mall at UBC. Probably the largest tree in town is between the wings of the Chemistry Bldg along Main Mall at UBC.

Sorbus aucuparia L.
European Mountain-Ash or Rowan

This small, deciduous European tree is a familiar part of our gardens and parks; it also escapes into the edges of forests and weedy places so that it is now well-established as a part of our local flora. The trees have pinnately-compound leaves, dark green above and paler beneath, with a row of teeth along the margins. Large flat-topped clusters of small white flowers in late spring are followed by great hanging clusters of bright orange fruits in the autumn and into winter, unless the birds devour them.

Among the many street plantings are those along the south side of Whyte Ave from Arbutus St for a half-block east, along 8th Ave and 9th Ave between Sasamat St and Tolmie St, along 12th Ave between Tolmie St and Blanca St, along 30th Ave between Highbury St and Dunbar St, large ones on 36th Ave between Blenheim St and Collingwood St, along the west side of Western Parkway between Chancellor Blvd and University Blvd (alternating with various *Crataegus* cultivars), and a large individual on the north side of 10th Ave between Balaclava St and Trutch St.

'Fastigiata' – This is the narrow, upright form, like a small Lombardy Poplar. There are two individuals on the north side of 8th Ave between Spruce St and Alder St, one on the north side of 4th Ave between Trutch St and Balaclava St, and a large one with a slightly broader crown in the Old Arboretum at UBC.

Sorbus decora (Sarg.) C.K. Schneid.
Showy Mountain-Ash

An eastern North American small tree that is rarely cultivated outside botanical collections. The leaves are shinier than those of the European Mountain-Ash and have more leaflets (usually 15) that are longer, more pointed, and have many more teeth. Large, dark brown winter buds are sticky. The fruits are 6–10 mm across, bright orange-red in large clusters, and are usually eaten by birds as soon as they are ripe. There is sometimes good orange-red autumn leaf colour.

There is a group of three trees on the west side of Queen Elizabeth Park that are all very similar in habit and are probably grafted on a straight trunk.

Sorbus hupehensis C.K. Schneid.
Hupeh Mountain-Ash

This lovely small tree from the province of Hupeh in China grows to about 10 m tall and is attractive for much of the year. The young twigs are red and bear-toothed with pinnately-compound, sea green leaves on red petioles. Flat-topped, white flower clusters, typical of the genus, are produced in late spring. These are followed by white or soft pink fruits produced in great abundance. Birds do not usually eat the fruits, as they do the orange-fruiting species, so the fruits may remain on the trees well into winter.

There are trees on the north side of the Stanley Park Pitch & Putt Golf Course, and several among the heathers and in the Sino-Himalayan Garden at VanDusen Botanical Garden and in the Asian Garden at UBC Botanical Garden.

'Pink Pagoda' – This new cultivar was named by UBC Botanical Garden in 1988. It has fruits that begin dark pink, fade to almost pure white by late autumn, and change to pink again in winter before they drop. It is a very superior form to most of the ones cultivated. The cultivar is just being promoted and distributed by the Botanical Garden and should soon be seen around the city. The original tree is just inside the Moon Gate (to the south) of the Asian Garden at UBC Botanical Garden.

Sorbus × thuringiaca (Ilse) Fritsch
(*Sorbus aria* [L.] Crantz × *Sorbus aucuparia* L.)
Hybrid Mountain-Ash

This garden hybrid has variable leaves that fall between the simple lobed leaves of the first parent and the pinnately-compound of the second. The hybrid usually has lobed leaves, except for the basal pairs which are often completely cut to the mid-rib. The leaves are narrowed to a distinctive point at the tip. The fruits are smaller and darker red than those of *Sorbus aucuparia*.

It is not often seen in cultivation in Vancouver. There are two trees on the NE corner of 5th Ave and Balaclava St, one on the southern corner of the Stanley Park Lawn Bowling area, one in the Sorbus Collection at VanDusen Botanical Garden, and a row of four trees at a service station on the SW corner of 41st Ave and Knight St (with one tree of each of the parents).

Stranvaesia davidiana Decne.
Stranvaesia

A very cotoneaster-like evergreen shrub or small tree growing to at least 7 m tall in slightly warmer climates to the south of us, especially in the San Francisco area. It is native to western China and is relatively unknown in our area outside parks and public gardens. The bright green leaves (to 12 cm long) have wavy margins, a characteristic that separates it from the similar, but very flat-leaved, cotoneasters. Small white flowers in flat-topped, drooping clusters in summer are followed by dull red fruits. The fruits are smaller than those of most of the tree-form cotoneasters.

There are some large tree forms in the Asian Garden at UBC Botanical Garden, and several in Stanley Park, including individuals on the north side of the Fish House Restaurant, by the Children's Zoo ticket office and hanging over the nearby pool (there are some large *Cotoneaster × watereri* nearby for comparison), and a large multi-trunked form on a bank above the parking area west of the Dining Pavilion.

Rutaceae – Rue Family

Phellodendron amurense Rupr.
Amur Cork Tree

This Chinese deciduous tree grows to about 20 m tall and is generally nondescript until it becomes very old and develops attractive, furrowed grey bark. There are no trees in Vancouver old enough, as yet, to show this characteristic. The shiny, pinnately-compound leaves are composed of 7–11 ovate leaflets, looking very much like an ash. Clusters of small green flowers are produced on the ends of new growth in late spring to early summer and are followed by round green berries that turn black in September, then shrivel, resembling raisins, and remain on the branches over winter. There are separate male and female trees.

There are four fairly large trees on the east side of Queen Elizabeth Park along the road up to the Bloedel Conservatory and one west of the Rose Garden on the west side of the park, two trees around the Pitch & Putt Golf Course in Stanley Park, and street plantings with cherry trees on both sides of Alberta St between 45th and 46th Ave and on the south side of 23rd Ave from Ash St to Tupper St. •79

Ptelea trifoliata L.
Hop Tree

This attractive eastern North American, deciduous, shrubby tree grows to about 8 m tall, with leaves that have 3 leaflets, smooth margins, and long pointed tips. Small yellow-green flowers are borne in round clusters, 5–8 cm across, at the ends of new growth in mid-June. The flowers are followed by flat, wafer-like fruits with a seed in the middle of each (samaras, like those of the elms). Leaves, and especially immature fruits, have a strong aromatic smell when crushed, typical of citrus leaves and many other members of this family.

There is a relatively large one on the hill east of the Seasons in the Park Restaurant in Queen Elizabeth Park, and one along the western end of the Rhododendron Walk in VanDusen Botanical Garden; the largest one in the city (larger than most in their native habitat) is on the north side of the Frederic Wood Theatre on Crescent Rd at UBC.

Salicaceae – Willow Family

Populus – Poplars and Aspens

Although members of the willow family, the poplars have thick, broad leaves and do not look very much like the willows. The leaves are often very triangular in outline, and the petioles of some species are flattened, allowing the leaves to flutter in the breeze. The flowers are produced in drooping catkins and the white, fluffy seeds resemble those of the related willows. Many have pointed, resinous winter buds and leaves that are strongly aromatic as they unfurl in the spring. Poplars are a promiscuous lot, with many hybrids produced between the species, both in nature and in cultivation.

Populus leaves
1 *P. alba*
2 *P. alba*
3 *P. balsamifera*
4 *P. canescens*
5 *P. grandidentata*
6 *P. × canadensis*
7 *P. nigra*
8 *P. tremuloides*

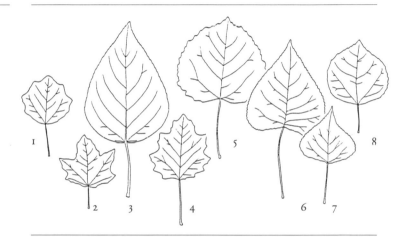

Populus alba L.
White Poplar

The silvery-white, lobed and toothed leaves cause this tree to be confused with the Silver Maple. This European native grows to 30 m tall and is cultivated frequently, and often naturalized, in climates colder than ours. The upper surface is dark grey-green, and the undersurface silvery due to a dense coat of white hairs, especially prominent on young leaves and twigs. On windy days, the silvery colour is very obvious. Root sprouts are sent up all around the tree, forming groves – an undesirable characteristic for gardens.

It is surprisingly uncommon in the city. There are individuals in a garden on the east side of John Hendry Park just north of 19th Ave, in the lane between 10th Ave and 11th Ave between Burrard St and Cypress St, and on the sw corner of MacMillan Bldg at UBC; there are rows on the east side of Victoria Dr just

north of Venables St, and on the lane on the east side of
Balaclava St just north of 41st Ave. The largest must be a tree on
the NE corner of Cordova St and Dunlevy Ave.

Populus balsamifera L. ssp. *trichocarpa* (Torr. & Gray) Brayshaw (*Populus trichocarpa* Torr. & Gray)
Black Cottonwood or Western Cottonwood

This is one of our most common native deciduous trees,
typically found on river bottoms, where it becomes very large
(to 50 m tall), but also in drier forests. It reseeds readily, so
seedlings may be found in almost any habitat. The tree is
sometimes left in gardens or intentionally cultivated. The over-
wintering leaf buds are large, long-pointed, and sticky to the
touch. In spring, the young unfurling buds and pale green
leaves have a strong smell, pleasing to most people. The flowers
are borne in long pendulous catkins, later producing fluffy
white wind-borne seeds. Large, thick leaves turn a bright golden
yellow in the late autumn.

It is a common native tree of forests in Pacific Spirit Regional
Park and Stanley Park. There are several large trees on the east
side of Queen Elizabeth Park, west of Ontario St and south of
33rd Ave, a number of large ones along 4th Ave below Jericho
Hill School, and a large individual on the SE corner of 57th Ave
and Granville St.

Populus canescens (Ait.) Sm.
Gray Poplar

This tree is thought to be a hybrid between the White Poplar
(*Populus alba*) and Black Poplar (*Populus nigra*). The leaves are
grey beneath, rather than the pure white of White Poplar, and
they usually become greener during the summer, with only the
newest growth on the tips retaining the grey colour. The leaves
are broadly triangular to diamond-shaped and have a few large,
irregularly rounded teeth. The typical form has rather upright
branches with sinuous drooping tips.

There are nice old individuals in Queen Elizabeth Park, a large
one in Stanley Park on the south side of Lost Lagoon at the foot
of Haro St (west of the tennis courts), and two on the NE side
of Jericho Beach Park.

'Pyramidalis' – This cultivar is the more common form found
here. There are often root sprouts, forming colonies of smaller
plants around the parent trees. The tree resembles a grey-leaved

Lombardy Poplar with whiter leaves and smooth, pale grey-white bark. There are two old specimens at the Justice Institute of BC on 4th Ave, one on the north side of York Ave between Arbutus St and Yew St, a large one on the SE side of Beatty St between Dunsmuir St and Georgia St, two young trees against a building on the SW corner of 8th Ave and Willow St, and a group on an island in the western end of Lost Lagoon in Stanley Park.

Populus grandidentata Michx.
Large-Toothed Aspen

This large, eastern North American poplar grows from Quebec and Ontario southward and west to extreme southeastern Saskatchewan, and is very rarely cultivated. The large, broad leaves have very prominent rounded teeth, which give it both the scientific and the common names. The leaves are borne on flattened petioles, thus allowing the leaves to quake in the breeze, and are slightly downy beneath when unfurling in the spring. The bark is smooth and pale grey-green on young trees, resembling that of the Quaking Aspen, but it becomes dark grey and furrowed on old trunks.

The only specimen known in the city is a 3-trunked tree in the Old Arboretum on the UBC campus, but it has been suggested that this tree is actually a hybrid between this species and Quaking Aspen.

Populus nigra L. 'Italica'
Lombardy Poplar

This well-known columnar form of the Black Poplar is a male clone that originated from a single tree in Lombardy, northern Italy, early in the eighteenth century. It may be 30 m tall at maturity and always remains very columnar. It is probably our best-known column-shaped tree. The leaves are almost triangular in outline, with flattened petioles.

It is abundant here, especially in parks. There is a large grove on the UBC campus east of the Agriculture Canada Research Station on NW Marine Dr and another between the Bus Loop and the Student Union Bldg, many west of Burrard St between English Bay and Beach Ave, two very large specimens on the SW corner of 15th Ave and Balsam St, and a row of large ones on the east side of Adera St between 54th Ave and 57th Ave.

Populus tremuloides Michx.
Quaking Aspen

This is an abundant tree in the dry interior and northern parts of our province and southward, east of the Cascade Mountains. It is not common on the wet, west coast. It tends to be found locally only on dry rocky outcrops just above the sea, such as in Lighthouse Park in West Vancouver. The green-white bark is distinctive from afar, even on mature trees. The flattened petiole of the leaves allows them to quiver in the breeze, thus the common name. Furry grey catkins are produced in February to March. The brilliant gold colour of the turning leaves of this tree is one of the distinctive features of autumn in the interior regions of western North America.

It is rarely planted here. There is a nice grove (probably native) of relatively large specimens at Spanish Banks on the west side of Trimble St, a young grove on the south side of Granville Island, and one specimen in the Old Arboretum at UBC. There are a few wild specimens in Pacific Spirit Regional Park at the western end of King Edward Ave.

Populus hybrids

There is a group of fast-growing hybrid poplars, mostly *Populus* × *canadensis* (*Populus deltoides* × *Populus nigra*), that has been cultivated and promoted in recent years for its very rapid growth. There are also several other species involved in some of the hybrids. The trees are difficult to characterize because of their complex background. A locally cultivated poplar may be one of these hybrids if it does not seem to fit any of the species descriptions, and especially if it is a young tree.

There are several trees on the south side of Broadway between Penticton St and Slocan St (on the grounds of the Vancouver Technical Secondary School), a row on the north side of the lane between 12th Ave and 13th Ave and between Oak St and Laurel St, and a group of trees behind Buchanan Bldg on the UBC campus which also seem to be hybrids between several species. There is a nice golden cultivar, 'Aurea,' just at the entrance to VanDusen Botanical Garden.

Salix – Willows

Willows are often the dominant deciduous tree and shrub component of high latitudes, often in wet places. A few are even small ground-hugging arctic and alpine plants. The fuzzy catkins, usually either male or female, are characteristic of many species. Large buds with a single bud scale covering the immature catkins are a distinguishing winter feature. On many species, the leaves are lance-shaped, a familiar enough shape that slender leaves of many other plants are often described as being 'willow-like,' but some willows do have much broader leaves. Many species have glossy, colourful, pliable twigs, used in weaving baskets and the like. There are some 300 wild species, a few of which, especially the weeping willows, are familiar as ornamentals.

Salix leaves

1 *S. alba*
2 *S. chrysocoma*
3 *S. lasiandra*
4 *S. matsudana*
5 *S. nigra*
6 *S. pentandra*
7 *S. scouleriana*

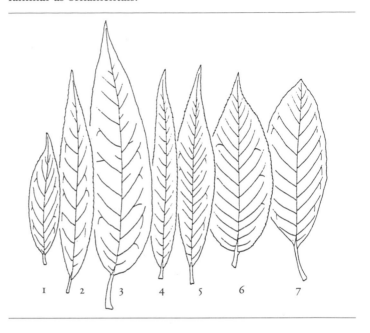

Salix alba L.
White Willow

This is a common hardy tree of Eurasia and North Africa, growing to 25 m tall. The lanceolate leaves, to 10 cm long, are finely toothed and have a white-silky pubescence beneath, and sometimes above also. The twigs are often golden-coloured in winter and early spring. Catkins are yellow-green, appearing with the new leaves. There are a number of different forms occurring naturally and in cultivation. The pliable twigs have long been used for making baskets.

It is not common here. There is a row on the south side of the Prospect Point Picnic Area in Stanley Park, and several nice trees in Jericho Beach Park that are visible from a long distance (particularly in winter) because of the yellow erect twigs. They may also be compared to the more common, but poor, Black Willow.

'Argentea' – This form is a round-topped tree with very silvery leaves that remain silky on both sides, even when mature. It appears quite silvery at a distance. The only specimen seen is a nice one on the south side of the lake in Jericho Beach Park.

Salix × *chrysocoma* Dode (*Salix alba* L. 'Tristis')
Golden Weeping Willow

Weeping willows are probably among the best known of all our cultivated trees, even to those people who recognize few trees. The graceful weeping habit, golden yellow twigs in winter, very pale green leaves in early spring, bright green leaves in summer, and golden late autumn colour make this large tree a favourite for parks and large gardens. The slender leaves have small teeth around the edges and are very sparsely hairy on the lower surface when young. Pale yellow-green catkins are obscured as the new leaves unfurl. There is much discussion and confusion around the correct identity and origin of the various weeping willows. *Salix babylonica* has much darker twigs and larger leaves, and apparently is not grown this far north, although it is the common form in more southerly climates. The Golden Weeping Willow is thought to be of hybrid origin, occurring long ago, probably in France, and goes under a number of different scientific names in various books, although *Salix* × *chrysocoma* seems to be the accepted name for the forms found in the Vancouver area. The trees prefer wet soil and are at their best around lakes and ponds, where the graceful branches reflect in the water.

Among the large specimens in the city are those around Lost Lagoon in Stanley Park, several large specimens on the SE side of Trout Lake in John Hendry Park, three large ones on the SW corner of 20th Ave and Heather St, several in Memorial South Park on 41st Ave at Prince Albert St, several around the Langara Campus of Vancouver Community College along 49th Ave from Columbia St to Ontario St, and a large individual on the SE corner of 16th Ave and Trafalgar St. •80

Salix lasiandra Benth.
Pacific Willow

This upright or spreading (not weeping) tree is our largest native willow and grows to about 12 m tall, although it often remains a small bushy tree. It is found in wet places, especially along forest edges, along the Pacific Coast from Alaska to California. The long, pointed leaves are 6–15 cm long and 1–3

cm wide, among the largest of our local willows. They are shiny green above and glaucous grey beneath. There is a pair of rounded stipules at the bases of the leaves, but these drop early in the season. The leaves of the flowering shoots are only about half the size of those on the vigorous young vegetative shoots. The catkins, which appear as the young leaves are unfurling, are pale yellow-green, 3–10 cm long, and lack any soft hairs such as those seen on the pussy willows.

It is a widespread tree in British Columbia, but not at all common in the city, although it probably was at one time. It is rarely, if ever, cultivated. Wild individuals may be seen in Stanley Park and Pacific Spirit Regional Park, especially along 16th Ave west of Imperial Dr and at the western end of King Edward Ave. There are also several trees in the open area on the sw corner of Trimble St and Belmont Ave, with several of the much more common Scouler's Willow, which has broad leaves with blunt tips that are dull green above and soft beneath.

Salix matsudana G. Koidz. 'Tortuosa'
Twisted Twig Willow or Corkscrew Willow

This hardy willow is native to northeastern Asia and is usually seen in gardens as the cultivar 'Tortuosa.' It has relatively narrow upright growth and sinuous twigs that bend in the opposite direction to each leaf. Most noticeable in winter, the twigs give the tree a distinctive look that is admired by flower arrangers. Unfortunately, the trees suffer from fungal diseases that result in many dead twigs and limbs, and they look quite miserable in our area.

It is not very common in cultivation outside parks and botanical gardens. There is a relatively large tree on 5th Ave just west of Burrard St, one on the NW corner of 16th Ave and Susumat St, a relatively large one and a small, poor specimen on the east side of Arbutus St just north of Valley Dr, one on the east side of Camosun between 37th Ave and 38th Ave, one in the Old Arboretum at UBC, and several in the UBC Botanical Garden and at VanDusen Botanical Garden.

Salix nigra Marsh.
Black Willow

A large willow native to eastern North America, this tree usually looks very scruffy in our climate. It is infested with various fungal diseases that cause twig dieback and the leaves to turn brown and drop early in the season. The leaves are lanceolate,

to about 12 cm long, dark green above and pale green beneath, and have few or no hairs. The catkins appear with the new leaves. The name comes from the very dark brown or black bark.

It is a fairly common, but poor, tree in parks in Vancouver. There is a row on the north side of 45th Ave between Ross St and Elgin St, a group on the south side of Memorial South Park, a long row on the north side of Trout Lake on the north side of John Hendry Park from Templeton Dr west for two blocks, a group on the east side of Queen Elizabeth Park west of Ontario St between 33rd Ave and 37th Ave, and a row along MacKenzie St between 16th Ave and 18th Ave on the east side of Carnarvon Park.

Salix pentandra L.
Bay Willow or Laurel Willow

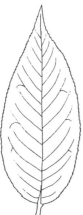

This is a Eurasian tree with unusually broad, thick leaves for a willow. They are up to about 8 cm long and 4 cm wide, with fine teeth along the margins, a slender pointed tip, and a shiny upper surface that looks as if it has been polished. It is very late to bud and leaf out in spring. Pale green catkins are borne after the leaves emerge. The trees grow to about 10 m tall and form a round-topped outline at maturity.

The only tree in Vancouver seems to be a large female just west of the Vancouver School of Theology along Iona Dr just south of Chancellor Blvd at UBC.

Salix scouleriana Barratt in Hook.
Scouler's Willow

This is the only common native willow throughout our area, often found in the edges of forests. It is often a multi-trunked large shrub, but may become a fairly large tree. The leaves are a bit wider than what one thinks of as being typically willow-like, with a dull upper surface and dense, soft white to grey or rusty hairs beneath. It is usually overlooked when in leaf, but is quite visible at a distance in late winter and early spring when the silvery catkins begin to emerge. It is one of the earliest trees to show signs of life, often as early as January during mild winters. There are separate male and female trees, but they both show silvery catkins at first. Later the males become yellow and more showy as they release their pollen. The females remain grey when in full flower.

This willow may be seen throughout natural areas of our city, in Stanley Park, Pacific Spirit Regional Park, and in many of our parks where they have been allowed to grow. It is not often cultivated, but there is a large multi-trunked male tree in a yard on the NE corner of 17th Ave and Ash St, one on the west side of Yew St between 13th Ave and 14th Ave, and several on the SW corner of Trimble St and Belmont Ave (with *Salix lasiandra*).

Sapindaceae – Soapberry Family

Koelreuteria paniculata Laxm.
Golden-Rain Tree

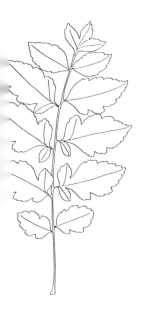

This is one of the few members of a large tropical family that is hardy this far north. Golden-Rain Tree is a small, rounded tree growing to about 10 m tall and native to China, Japan, and Korea. It has large compound leaves with variably toothed leaflets. The broad clusters of small golden yellow flowers are showy in summer, hence the common name. The flowers are followed by 3-angled, bladder-like, green fruits, about 3 cm long, which become papery and turn a rusty gold in late autumn and winter. We are near the northern limit for successful growth of this tree.

There are a very few specimens in collections in Vancouver, but it seems to do relatively well and should be grown more often. There is a nice small tree on the north side of 31st Ave between Blenheim St and Collingwood St, several trees just south of the Heather Garden and others just west of the pine woods in VanDusen Botanical Garden, and some young ones in the Asian Garden and around the new entrance to UBC Botanical Garden. A row on the south side of East Mall Classroom Block on the UBC campus had the largest ones in the city, but these were cut a few years ago for expansion of a parking lot.

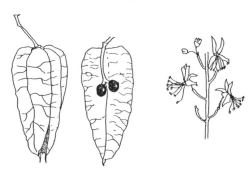

Simaroubaceae – Quassia Family

Ailanthus altissima Swingle
Tree-of-Heaven

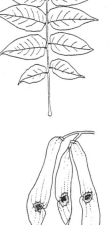

This native of China is one of the most tolerant of all trees to city conditions. It is often seen in eastern North American cities, where it is considered a weed tree because it reseeds freely and has become widely naturalized. In Vancouver it does not usually reseed, is not weedy, and becomes an attractive specimen tree, if given time. The large pinnately-compound leaves give a somewhat tropical effect. The leaflets often have one or more irregular lobes at the base, a characteristic that may be used to separate Tree-of-Heaven from walnuts (*Juglans*) or other trees with large pinnately-compound leaves. There are separate male and female trees. Both have small yellow-green flowers borne in large clusters in midsummer. Females bear distinctive, yellow to orange or red, winged fruits that turn brown and hang on the trees over winter.

It is not a common tree in Vancouver. There are several large specimens around the city, mostly females. The tallest trees are probably two in Stanley Park, one among a grove of old trees near the Rose Garden (south of the Service Yard) and a second near the end of the pedestrian walkway. Although not the tallest, the most massive and attractive specimen is a very beautiful female tree in the small triangular park bounded by Kingsway, Fraser St, and 15th Ave. There are female trees on the east side of Dunbar St between 4th Ave and 5th Ave, on the north side of 4th Ave between Balaclava St and Trutch St, several on the north side of 8th Ave east of Alder St, and one in the Old Arboretum on the UBC campus.

Styracaceae – Styrax Family

Halesia tetraptera Ellis
(*Halesia carolina* L., *Halesia monticola* [Rehd.] Sarg.)
Silver-Bell Tree

There is a great deal of confusion as to the correct name of this very ornamental tree. Most of the standard references list it as *Halesia carolina* or *Halesia monticola,* or one as a variety of the other. Most recent authorities on the genus consider the correct name to be *Halesia tetraptera,* meaning 4-winged, referring to

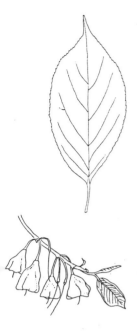

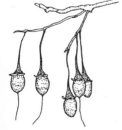

the fruits. It is a small deciduous tree native to the southeastern United States, reaching 35 m tall in its native habitat, but usually seen in cultivation as an irregularly arching, multi-trunked, smaller tree. The oval leaves are 10–15 cm long. The tree is especially showy in May when the limbs are covered in drooping, white, bell-like flowers, 2–3 cm long. Later in the year, there are attractive, angular, coppery seed capsules that hang from the bare winter branches, often lasting until the next year's flowers are open.

It is infrequently cultivated here. There are two large multi-trunked specimens in the Old Arboretum at UBC, two others just north of the Anthropology and Sociology Bldg between NW Marine Dr and Cecil Green Park Rd at UBC, a nice, large specimen on the SW corner and a smaller one on the NW corner of the tennis courts at Tatlow Park (Macdonald St and Point Grey Rd), and one west of the Quarry Garden in Queen Elizabeth Park.

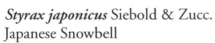

Styrax japonicus Siebold & Zucc.
Japanese Snowbell

A small deciduous tree from China and Japan that grows to about 10 m tall and is cultivated in gardens for its masses of white flowers borne in early summer, usually June. The small leaves are elliptic in outline. The pendulous flowers, about 2 cm long, are 5-petalled and bell-shaped, with bright yellow stamens, and are best viewed from beneath the branches. They are followed by hard, green fruits which later turn brown and hang from the branches throughout the autumn and winter. They are not as showy as some of our ornamental trees, but they are charming when in flower.

There are two large specimens in The Crescent, three at Stanley Park Zoo by the seal cages, one on the north side of the Graduate Student Centre at UBC, three small trees on the north side of 12th Ave between Blanca St and Tolmie St, a row on

Hudson St from 40th Ave to 41st Ave, one on the east side of Queen Elizabeth Park by the waterfall, and a number of specimens (as well as other unusual species of *Styrax*) in the Asian Garden of UBC Botanical Garden.

Styrax obassia Siebold & Zucc.
Fragrant Snowbell

This snowbell is less common than the previous species. It is easily distinguished by its large oval leaves, to 25 cm long, with a soft pubescence beneath. The fragrant, white, bell-like flowers are borne on long racemes in early June, usually about two weeks earlier than *Styrax japonicus*. It ultimately becomes a tree growing to about 10 m tall.

There are several large specimens on the west side of Nitobe Memorial Japanese Garden and in the Asian Garden at UBC Botanical Garden, a large one in the Old Arboretum at UBC, four trees in Stanley Park north of Lagoon Dr at Barclay St and around the Pitch & Putt Golf Course, and three small trees on the south side of Angus Dr just west of Granville St. •81

Tamaricaceae – Tamarisk Family

Tamarix parviflora DC.
Tamarisk or Salt Cedar

A large shrub or small, often multi-trunked, tree native to southeastern Europe, and now naturalized along streams and in saline soils throughout many parts of the world. When not in flower, the small scale-like leaves and wispy, green branches give the appearance of some kind of conifer, rather than a flowering plant, but the twigs are transformed completely when they are covered with thousands of tiny, 4-petalled, soft pink flowers in May. The flowers are only 2–3 mm wide but are borne in dense racemes, 3–4 cm long, covering the twigs. There are several other *Tamarix* species, with 5-petalled flowers that are very similar to this species, and they, too, may be in cultivation here.

There are a few, small, shrubby plants around the city, and a few that definitely qualify as small tree forms. The larger ones are located on the east side of Alma St between 2nd and 3rd Ave and on the NW corner of 45th Ave and Sherbrooke St; there are two on the NE corner of 33rd Ave and Pine Cres, one on the west side of Balaclava St between 37th Ave and 38th Ave, and one on the SE corner of Valley Dr and 22nd Ave.

Theaceae – Tea Family

Camellia japonica L.
Japanese Camellia

An evergreen shrub or small broad-leaved evergreen tree growing to 12 m tall in its native habitat in China and Japan. It is grown commonly in the warmer temperate parts of the world for its late winter or spring flowers of white to pink and red, and in a huge array of cultivars. It is usually seen as a dense shrub here, but will become a tree to several metres tall, especially if pruned into a tree form at a very early age. The typical wild forms have a few petals, usually 5, with a large cluster of bright yellow stamens in the centre, but most of the hundreds of cultivated varieties have fully double flowers with many petals and often lack stamens. In our area the white forms usually are unattractive because the finished flowers turn brown and hang on the bushes much too long. Many of the oldest specimens in the city are of one or a few similar cultivars with very formally double, carmine red flowers, but many old cultivars are not easily identified. The peak flowering for Japanese Camellias in Vancouver is usually in late March to early April.

Large tree-like specimens in Vancouver include those on the north side of 10th Ave at Laurel St, on the NE corner of 10th Ave and Discovery St, on the north side of King Edward Ave west of Manitoba St, some large specimens on the NE corner of 3rd and Alma St, on the SW corner of 3rd Ave and Cypress St, on both sides of Stephens St between 7th Ave and 8th Ave, in Nitobe Memorial Garden at UBC, and in the Camellia Collection at VanDusen Botanical Garden.

Stewartia pseudocamellia Maxim.
Japanese Stewartia

There are far too few of this lovely small tree in our area. It is, sadly, almost unknown outside botanical gardens and parks. It ultimately reaches about 15 m tall in its native Japan. The flowers are borne in June and thereafter, and sometimes there are a few later in the summer. They have 5 ruffled petals with a large cluster of orange-yellow stamens in the middle, and open to about 4 cm across. They resemble a small white Camellia (hence the name *pseudo*, meaning 'false,' and *camellia*). The leaves are rather nondescript, except that they are dull green above and shiny beneath, with fine teeth around the edges.

They often develop good autumn colours of oranges and reds. The bark becomes mottled and very attractive on older specimens. The pointed, 5-parted seed pods remain on the trees over winter and are also an attractive coppery brown.

There are specimens around the Stanley Park Pitch & Putt Golf Course, along the Rhododendron Walk in VanDusen Botanical Garden, in the Asian Garden of UBC Botanical Garden (with other *Stewartia* species), one north of MacMillan Bldg on Main Mall (west of the Barn Coffee Shop), and two small trees in front of an apartment on the west side of Spruce St between 13th Ave and 14th Ave. •82, 83

Tiliaceae – Linden Family

Tilia – Lindens, Limes, and Basswood

Lindens are among the most commonly planted street and park trees in temperate parts of the world. Most are very similar in leaf and flower characteristics and are not easily separated. The heart-shaped leaves usually have long petioles, uneven (oblique) leaf bases, and toothed margins. The very fragrant, creamy white flowers are borne in drooping, rounded or flat-topped clusters with a single long leaf-like bract at their base. In addition to the species treated here, there are a number of others in the Tilia Collection at VanDusen Botanical Garden.

Tilia americana L.
American Linden or Basswood

This large eastern North American linden grows to as much as 40 m tall and its leaves are among the largest of any lindens (10–20 cm long). The prominent teeth around the leaves have long slender points. The leaves are dark green above, paler beneath, and have hairs along the lateral veins but not at the base of the midrib. Typical yellow, fragrant, drooping flowers are produced in July. The leaves are lost rather early in the autumn, turning pale or dull yellow before dropping.

The typical wild form is rarely cultivated here. The only specimens seen are a row of large trees on 13th Ave from Granville St to Fir St, several interplanted with *Tilia platyphyllos* in a row (planted by graduating classes) along the east side of the Geography Bldg between Main Mall and West Mall at UBC, and one in the Old Arboretum at UBC.

'Redmond' – There is some confusion as to the parentage of this cultivar that has become a popular street tree in recent years. It is sometimes considered a selection of the hybrid Crimean Linden (*Tilia* × *euchlora*), but it is more often assumed to be a selection of the American Linden. It has a dense, oval top (or broad tear-shape), tends to be very tidy-looking, and isn't infested by aphids as much as other lindens. There seem to be no old trees in the city, but a number of trees were planted along major streets during the 1970s. Among the many lengthy plantings are those along 4th Ave between Burrard St and Balsam St, along Broadway from Waterloo St to Larch St, along Main St from 16th Ave to 33rd Ave, and along Tyne St between 49th Ave and 54th Ave.

Tilia cordata Mill.
Little-Leaved Linden

Although variable in size, the leaves of this linden are the smallest of any of our cultivated species. They are often only 3–6 cm long and are glabrous above and below, except for tufts of brown hairs in the vein axils. Leaves are often glaucous beneath. The similar Large-Leaved Linden (*Tilia platyphyllos*) has distinctly soft-pubescent leaves both above and below, and it is common in Vancouver. Little-Leaved Linden is a European native growing to 30 m tall.

An abundantly planted tree in many cities (as in Toronto), it is surprisingly rare in Vancouver. Some trees may have been overlooked due to the similarity between this species and small-leaved individuals of *Tilia platyphyllos*. There is a row on Waterloo St between 12th Ave and 13th Ave, and in front of the Biomedical Research Centre (opposite the parkade) on the east side of Health Sciences Mall at UBC.

Tilia petiolaris DC.
Pendent Silver Linden

Drooping leaves and branches, long petioles, and very dark green leaves with silvery white undersides that shimmer in the breeze are definite characteristics of this rare, but beautiful, linden. The tree, which grows to about 25 m tall, is thought to be native to SE Europe and western Asia, although it is not known in the wild. The similar Silver Linden (*Tilia tomentosa*) has more erect branches and shorter petioles.

There is a lovely specimen on the sw side of Queen Elizabeth Park between the parking area off 37th Ave and the paved walkway to the Quarry Garden, and one in the Old Arboretum at UBC.

Tilia platyphyllos Scop.
Large-Leaved Linden

A common European linden, which, in spite of its name, has among the smallest leaves of any of our locally cultivated lindens. The best distinguishing characteristic is the soft pubescence on both sides of the leaves (especially when young), plus hair tufts in the vein axils beneath. The leaves have a definite softer feel and, although variable in size (usually smaller on older trees), are generally smaller than those of most other lindens.

The flowers are very noticeable in June. Unfortunately, the trees suffer from attacks of aphids and the resulting black sooty moulds coat the leaves by late summer, making them unsightly.

This is probably the most common large linden in the city, especially in parks. Among the many in the city are large specimens in The Crescent; several around the perennial beds, the Zoo, and at the end of the walkway (by the causeway) at the entrance to Stanley Park; a large one beside a European Beech in the SE corner of Tatlow Park at 3rd Ave west of Macdonald St; a row on the north side of Midlothian Ave from Dinmont Ave east to Clancy Loranger Way (opposite the north side of Queen Elizabeth Park); and a row along the south side of 16th Ave at Carnarvon Park between MacKenzie St and Carnarvon St.

'Laciniata' – This form has leaves that are very irregularly dissected and cut. It has a very graceful and delicate texture, and is also a very heavy flowerer. It is rare in the city. There is a nice specimen in Queen Elizabeth Park just west of the parking area on the west side off 33rd Ave, two on the west side of Cambie St at 31st Ave, a relatively large one in a garden on the west side of Knight St just north of 33rd Ave, one in a row of Sycamore Maples on the north side of 15th Ave between Blanca St and Tolmie St, and one on the NE corner of 49th Ave and Butler St (at the end of a long row of typical wild forms).

Tilia tomentosa Moench
Silver Linden

Very similar to the Pendent Silver Linden (*Tilia petiolaris*), the Silver Linden also has leaves that are dark green above and white below, but it has more erect branches and shorter petioles. The white on the underside of the leaves comes from a very dense mat of very short silvery white hairs. Fragrant, creamy yellow flowers are usually out in late July. The tree, native to SE Europe and western Asia, grows to about 30 m tall.

It is not common here. The only trees seen in the city are two west of the tennis courts in Almond Park on 12th Ave between Dunbar St and Alma St, one on the north side of Cedar Cres east of Pine St, a very large one just south of the Stanley Park Service Yard (near the Rose Garden), a relatively large one on each side of Quebec St just north of 29th Ave, two on the east side of Yukon St between 13th Ave and 14th Ave, and a small one in the lawn area near the Main Garden Centre of UBC Botanical Garden.

Ulmaceae – Elm Family

Celtis occidentalis L.
Hackberry

This common deciduous tree grows to 40 m or more in eastern North America, but is often seen as a smaller tree along roadsides, in fields, and in the edges of forests. The 5–12 cm long leaves have oblique (uneven) bases, like those of most of the related elms. Hackberry leaves tend to be paler green, thinner, and less-toothed than elm leaves. Tiny yellow-green flowers appear in April or May, just as the leaves are emerging, and are followed by small, hard, orange-red to purple fruits which remain visible, hanging like tiny cherries from the bare winter branches. The bark is pale grey and smooth, except for small prominent, but irregular, ridges.

It is rare in Vancouver where the trees probably do not get enough extended summer heat to do well. The only ones seen are a single-trunked tree and a 3-trunked specimen in the Old Arboretum at UBC, and three young ones in the Eastern North American Section at VanDusen Botanical Garden.

Ulmus – Elms

This is a genus of only about eighteen species from temperate parts of the Northern Hemisphere, but it is very difficult to separate many of the species. The growth habit, especially in cultivation, and leaf size and shape are quite variable. One of the difficulties is that both the fruits and leaves are needed for positive identification, and the fruits are mostly borne in the

Ulmus leaves and fruits

1 *U. americana*
2 *U. carpinifolia*
3 *U. glabra*
4 *U. glabra*
5 *U. × hollandica*
6 *U. laevis*
7 *U. procera*
8 *U. pumila*

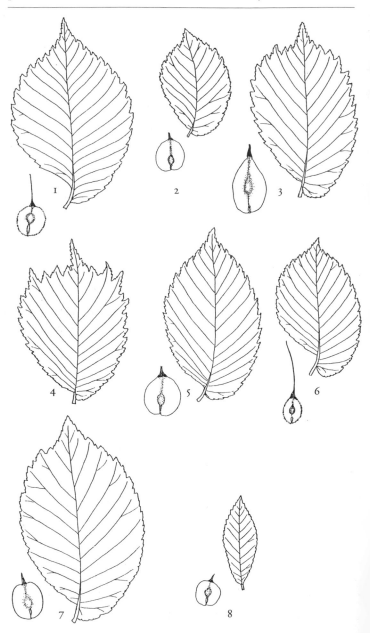

spring before the leaves are produced, or at least before they are mature. Most elms have oval or oblong leaves with a double row of marginal teeth. The petioles are very short and the leaves are often roughly hairy. Elms have two very distinctive characteristics. First, the leaves of most species have very uneven (oblique) leaf bases. Second, all elms have fruits (samaras) that are like a green or brown wafer, with a single seed in the middle. These may be produced in great abundance in the spring. Small green or brown flowers are insignificant and usually go unnoticed. A number of elms have a very distinctive vase-shaped habit when they are mature.

Ulmus americana L.
American Elm

This was one of the magnificent deciduous trees of eastern North America, growing to 35 m tall, and widely planted as a street and park tree until so many were killed by the introduced Dutch Elm Disease. Fortunately, the disease, or its bark beetle carrier, is not in the Pacific Northwest. There are many beautiful old specimens, with their characteristic vase-shaped crown, around the city. The small samaras with the seed in the middle appear in May and early June and have distinctive ciliate margins. The notch in the tip of the samara extends to the seed, a characteristic that may be used to separate it from the similar *Ulmus laevis*. The leaves of American Elm are generally larger, glabrous beneath, and are usually widest about the middle, which also helps to separate it from the latter species.

Among the good specimens in the city are several large and variably shaped trees along University Blvd between Main Mall and Wesbrook Mall at UBC, and an old tree in front of the Canadian National Railways Station between Station St and Main St; there are also street plantings on the south side of 10th Ave from Oak St to Laurel St, on 18th Ave between East Blvd and Cypress St, and on 18th Ave between Dunbar St and Collingwood St. •84

Ulmus carpinifolia Ruppius ex Suckow
Smooth-Leaved Elm

This is a relatively small-leaved elm from Europe, North Africa, and western Asia, growing to 30 m tall. It has been cultivated for centuries and has many selected clones. In our area, the wild forms are generally fairly small, sparse trees. It may be distinguished from several other similar species by a combination of

the following characteristics: glabrous young twigs; small leaves (4–10 cm long) which are glabrous and smooth to the touch on the upper surface; and samaras with the seed close to the tip.

There is a row of trees in the median of Beach Ave under the north end of the Granville Bridge, a group on the NW corner of Arbutus St and Cornwall Ave in Kitsilano Beach Park, a group in the NE corner of John Hendry Park, two large trees just east of the statue of Robert Burns near the entrance to Stanley Park, and two trees in a row of several *Ulmus* species to the east of Ceperley Picnic Area in Stanley Park.

'Sarniensis' – This is a very distinctive cultivar with a narrow pyramidal habit that is more commonly cultivated in Vancouver than the species. There are many around the parking lot of the Langara Campus of Vancouver Community College on the south side of 49th Ave east of Cambie St and there are street plantings along 10th Ave from Maple St to Vine St, along 23rd Ave from Oak St to Laurel St, and along Richelieu Ave west of Oak St.

Ulmus glabra Huds.
Scotch Elm or Wych Elm

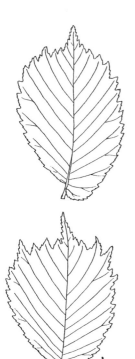

This large European native, growing to 40 m tall, is one of our easier elms to identify with its large, thick leaves that are often partially 3-pointed at the tip, especially on young vigorous branches and trunk sprouts. Leaf size and shape may vary greatly on the same twig. The leaves have a very short petiole, which is almost covered by the flaring, uneven leaf base. The name *glabra* refers to the smooth bark on young trees, certainly not to the leaves which are very rough, like sandpaper, on the upper surface and softer hairy beneath. The samaras are also larger (up to 2.5 cm long) than those of other local elms. It is our only local elm with the seed in the centre of the fringeless samara.

The wild form of the tree is not very common here. There is a street planting along 17th Ave between East Blvd and Cedar Cres, one tree in a row of Dutch Elms in Kitsilano Beach Park extending west along Arbutus St from the end of Whyte Ave, and a large tree (with an *Ulmus carpinifolia*) near the wading pool in Clarke Park at Commercial Dr and 14th Ave.

'Camperdownii,' Camperdown Elm – This is a very distinctive, drooping, mushroom-shaped small tree that is a familiar sight in parks and gardens throughout the temperate regions of the world. The leaves are larger than those of the species. There are a number of nice specimens around the city, including those in front of the Justice Institute of British Columbia on the north side of 4th Ave just east of NW Marine Dr, at the NE end of Lost Lagoon in Stanley Park near the underpass of the causeway, on the SW corner of 15th Ave and Balsam St, on the SE corner of 40th Ave and Marguerite St, on the SE corner of 41st Ave and Cartier St, and at UBC at Graham House (next to Cecil Green Park) and along the Main Mall at Sedgewick Library.

'**Lutescens**' – This golden leaved cultivar is very attractive, lighting up dull days. It is rare here, but there is a nice row of large specimens (alternating with Giant Sequoia) in the median of Cambie St between 27th Ave and 29th Ave. •85

Ulmus × *hollandica* Mill. (*Ulmus glabra* Huds. × *Ulmus carpinifolia* Ruppius ex Suckow)
Dutch Elm

These elms are a variable group of hybrids, within which there are a number of selected, named cultivars. The trees are more or less intermediate between the two parents, being variable in many characteristics. Hybrids may usually be identified by a combination of: relatively large leaves (6–12 cm long) which are smooth above; glabrous young twigs; and abundantly produced samaras which have the seed towards the tip. Dutch Elm is probably the most common elm cultivated in our area.

Specimens include a row of three trees (and one *Ulmus glabra*) west of Arbutus St at the foot of Whyte Ave (north of the tennis courts), a group of large trees east of the swimming pool in Kitsilano Beach Park, two trees in a row of mixed species east of Ceperley Picnic Area in Stanley Park, and street plantings along Churchill St between 41st Ave and 49th Ave and along both 13th Ave and 14th Ave between Pine St and Fir St. The largest specimen is probably one SW of the Main Library at UBC.

Ulmus laevis Pall.
European White Elm or Fluttering Elm

This elm is native to western Asia and central Europe and is not often cultivated in our area. It and the American Elm are the only elms cultivated locally that have samaras with hairs around the margins. The common name, Fluttering Elm, comes from

the long stalks (to 2 cm long) on which the samaras are borne, which allow them to flutter in the breeze. The stalks of American Elm are usually much shorter. *Ulmus laevis* may also be distinguished from American Elm by its soft, furry young shoots and leaves that are widest above the middle and soft furry beneath. Also, the samaras have a notch at the tip that does not extend to the seed.

There are street plantings on the south side of 14th Ave between Spruce St and Alder St, along 28th Ave and 29th Ave from Dunbar St to Highbury St, and on the south side of Marpole Ave just west of Granville St. There are possibly other specimens in the city, but they are easily confused with other elms, especially American Elm, unless seen in fruit.

Ulmus procera Salisb.
English Elm

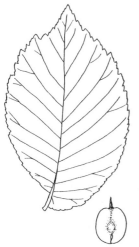

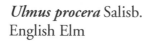

A common, large elm of Europe, it is not as common in Vancouver as several other similar species. The distinguishing features are soft, pubescent young twigs, relatively small leaves (5–9 cm long and petioles 4–8 mm long) that are rough above, and samaras with the seed near the tip. The only other elm cultivated locally with rough leaves is *Ulmus glabra,* but its leaves are larger, more often lobed toward the tip, and the petioles are shorter and usually hidden by the leaf bases. The young limbs sometimes have corky ridges.

Among the wild forms are two trees in a row with *Ulmus carpinifolium* and *Ulmus × hollandica* east of the Ceperley Picnic Area in Stanley Park, one on the NW corner of 41st Ave and Blenheim St, and one on the NE corner of 12th Ave and Spruce St.

'**Purpurascens**' – This cultivar has leaves that are dark bronze with darker purple veins, giving the tree a dark look, not as bright as the colour of Purple-Leaved Plums or Copper Beeches. The only two seen are tall slender trees on the SW corner of the Main Library (by the Map Library entrance) at UBC.

Ulmus pumila L.
Siberian Elm

The very hardy Siberian Elm is widely planted and, in some parts of North America, has become naturalized, and is even considered weedy. However, it is not common locally and does not seem to spread by seed. It may be confused with the similar,

but more desirable, Chinese Elm (*Ulmus parvifolia* Jacq.), which does not seem to be cultivated locally. Siberian Elm usually remains a small tree, ultimately growing to 10 m tall. The small leaves, 3–7 cm long with relatively even leaf bases and a simple row of marginal teeth, are unusual for elms, and unique among the ones cultivated locally. The samaras, produced in spring and falling by early summer, are smaller than other local elms. It is cultivated mainly because it is very cold-hardy and resistant to Dutch Elm Disease.

It is fairly common here. There is a street planting of moderately large trees along 11th Ave between Discovery St and Courtenay St, a row of trees near Pearson Hospital (George Pearson Centre) on the north side of 59th Ave from Heather St east for a half-block, a long street planting along 6th Ave between Nanaimo St and Grandview Hwy, and a row hanging over the sidewalk on the east side of Allison Rd at McMaster Rd in Point Grey.

Zelkova serrata (Thunb.) Mak.
Japanese Zelkova

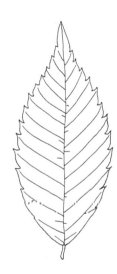

This large Japanese deciduous timber tree reaches 30 m or more tall and resembles the elms, which are closely related. The oval leaves, 5–12 cm long, have very prominent and regular teeth around the margins, and the leaf base is more regular than that of most elm leaves. There is often good golden autumn colour. It is rare in Vancouver, but deserves to be much more widely grown. The fruit is a small, hard, round drupe, very different from the familiar winged, circular samara of the elms.

It is rare here. The only ones found in the city are two relatively large specimens (one single-trunked and one multi-trunked) on the SE corner of the Maritime Museum, and two nice specimens at the entrance to the Floral Hall (by the parking lot) and three north of the Korean Pavilion at VanDusen Botanical Garden. •86

Verbenaceae – Verbena Family

Clerodendron trichotomum Thunb.
Glory-Bower Tree

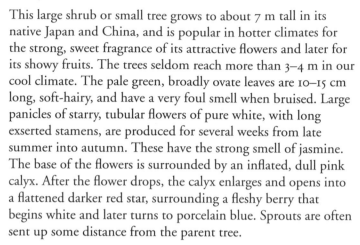

This large shrub or small tree grows to about 7 m tall in its native Japan and China, and is popular in hotter climates for the strong, sweet fragrance of its attractive flowers and later for its showy fruits. The trees seldom reach more than 3–4 m in our cool climate. The pale green, broadly ovate leaves are 10–15 cm long, soft-hairy, and have a very foul smell when bruised. Large panicles of starry, tubular flowers of pure white, with long exserted stamens, are produced for several weeks from late summer into autumn. These have the strong smell of jasmine. The base of the flowers is surrounded by an inflated, dull pink calyx. After the flower drops, the calyx enlarges and opens into a flattened darker red star, surrounding a fleshy berry that begins white and later turns to porcelain blue. Sprouts are often sent up some distance from the parent tree.

It is very rare in the city. There are three nice trees on the NW wall of the Fish House Restaurant in Stanley Park, two in the Sino-Himalayan Garden at VanDusen Botanical Garden, and a small one in the Winter Garden at UBC Botanical Garden.

flowering and
fruiting twigs

Glossary

anther	the end of the stamen (male flower part) which produces pollen
berry	a fruit that is fleshy at maturity, containing several seeds
blade	the broad, flattened part of a leaf
bract	a leafy structure, often below the flower
ciliate	with a fringe of fine hairs
columnar	a narrow, upright growth habit in which there is a central trunk maintained to the top of the tree, with shortened side branches (compare with **fastigiate**)
compound	a leaf composed of several leaflets (see **palmately-compound** and **pinnately-compound**)
conifer	a cone-bearing plant
cultivar	a cultivated variety
dichotomous	with pairs of equal branches
drupe	a single-seeded fruit that is fleshy at maturity
exfoliate	peeling away, often used in reference to bark
fasciated	broad, flattened, fan-shaped twig
fastigiate	an upright, slender growth habit in which there are a number of parallel stems or trunks (compare with columnar)
glabrous	lacking hairs, smooth
glaucous	with a white waxy coating on the surface of leaves, twigs or fruits, which may be wiped off easily
inflorescence	a flower cluster
keys	a common name for the fruits (samaras) of the maples
lenticels	corky 'breathing pores' in the bark, visible as raised bumps
needle	a slender, elongate leaf, especially of conifers
palmately-compound	a leaf made up of three or more leaflets attached at a common point
panicle	a multi-branched inflorescence
pedicel	the stalk of an individual flower
petal	usually the showy, coloured, leaf-like part of the flower

petiole | the stalk of a leaf which attaches the leaf blade to the twig

pinnately-compound | a leaf made up of one or more pairs of leaflets attached along the side of an axis (like a feather)

pistil | the female part of the flower

pollard | the practice of cutting limbs back to near the main trunk, giving the tree a dense, rounded (but unnatural) form

pubescent | covered with hairs, especially soft, fine ones

raceme | an inflorescence with a single long stem and short side branches, each bearing a single flower

rachis | the mid-rib or central 'stem' of a compound leaf, or the central stem of a cone

samara | a single-seeded, winged fruit, as found in maples, elms and ashes

sepal | usually a green leaf-like whorl of structures below the petals of a flower

sessile | lacking a petiole or stalk, a flower or leaf which is attached directly to the twig or stem

stamen | the male part of the flower

stipule | pairs of often leaf-like structures at the base of some leaves

References and Additional Reading

Arno, S.F. & R.P. Hammerly. 1977. *Northwest Trees.* Seattle: The Mountaineers. 222 pp.

Bailey Hortorium. 1976. *Hortus III (A Concise Dictionary of Plants Cultivated in the United States and Canada).* New York: Macmillan. 1,290 pp.

Bean, W.J. 1980. *Trees and Shrubs Hardy in the British Isles.* 8th ed. 4 vols plus supplement by D.L. Clarke (1988). London: John Murray. 3,990 pp.

Benson, D. & A. Cook. 1988. 'The Trees and Shrubs of Stanley Park: English Bay to Lost Lagoon.' In *The Natural History of Stanley Park,* edited by V. Schaefer & A. Chen. Vancouver: Discovery Press. Pp. 45–53

Chaster, G.H., D.W. Ross & W.H. Warren (J.W. Neill, ed.). 1988. *Trees of Greater Victoria: A Heritage.* Victoria, BC: Heritage Tree Book Society. 92 pp.

Dallimore, W. & A.B. Jackson. 1966. *A Handbook of Coniferae and Ginkgoaceae.* Rev. ed., by S.G. Harrison. London: Edward Arnold. 729 pp.

Den Ouden, P. & B.K. Boom. 1965. *Manual of Cultivated Conifers Hardy in the Cold and Warm-Temperate Zone.* The Hague, Netherlands: Martinus Nijhoff. 526 pp.

Dirr, M.A. 1983. *Manual of Woody Landscape Plants.* Rev. ed. Champaign, IL: Stipes Publishing Co. 826 pp.

Elias, T.S. 1987. *The Complete Trees of North America.* New York: Gramercy Publishing Co. 948 pp.

Godet, J. 1988. *Collins Photographic Key to the Trees of Britain and Northern Europe.* London: William Collins Sons & Co. 215 pp.

Harrison, C.R. 1975. *Ornamental Conifers.* New York: Hafner Press. 729 pp.

Hay, R. & P.M. Synge. 1986. *The Color Dictionary of Flowers and Plants.* 6th printing. New York: Crown Publishers. 584 pp.

Hillier & Sons. 1984. *Hillier's Manual of Cultivated Plants.* 5th. ed. London: David & Charles. 575 pp.

Hodel, D.R. 1988. *Exceptional Trees of Los Angeles.* Los Angeles: California Arboretum Foundation. 80 pp.

Hosie, R.C. 1979. *Native Trees of Canada.* 8th ed. Don Mills, ON: Fitzhenry & Whiteside. 380 pp.

Jacobson, A.L. 1989. *Trees of Seattle.* Seattle: Sasquatch Books. 432 pp.

Johnson, Hugh. 1973. *The International Book of Trees.* London: Mitchell Beazley Publishers. 288 pp.

Krüssmann, G. 1984. *Manual of Cultivated Broad-Leaved Trees and Shrubs.* 3 vols. Translated by Michael E. Epp. Portland, OR: Timber Press. 1,403 pp.

——. 1985. *Manual of Cultivated Conifers.* Translated by Michael E. Epp. Portland, OR: Timber Press. 361 pp.

Lawrence, E., ed. 1985. *The Illustrated Book of Trees and Shrubs.* New York: Gallery Books. 304 pp.

Macaboy, S. 1979. *What Tree is That?* London: Tiger Books International. 272 pp.

McMinn, H.E. & E. Maino. 1935. *An Illustrated Manual of Pacific Coast Trees.* Berkeley: University of California Press. 409 pp.

Miller, H. & S. Lamb. 1985. *Oaks of North America.* Happy Camp, CA: Naturegraph Publishers. 327 pp.

Mitchell, A. 1982. *A Field Guide to the Trees of Britain and Northern Europe.* London: William Collins & Sons. 288 pp.

——. 1987. *The Trees of North America.* New York: Facts on File Publications. 208 pp.

Muller, K.K., R.E. Broder & W. Beittel. 1974. *Trees of Santa Barbara.* Santa Barbara, CA: Santa Barbara Botanic Garden. 248 pp.

Phillips, R. 1978. *Trees of North America and Europe.* London: Pan Books. 223 pp.

Poor, J.M., ed. 1984. *Plants that Merit Attention. Volume I – Trees.* Portland, OR: Timber Press. pages un-numbered

Rehder, A. 1986. *Manual of Cultivated Trees and Shrubs.* Rev. ed. Portland, OR: Dioscorides Press. 996 pp.

Salmon, J.T. 1980. *The Native Trees of New Zealand.* Auckland: Reed Muthuen Publishers. 384 pp.

Taylor, R.L. & B. MacBryde. 1977. *Vascular Plants of British Columbia: A Descriptive Resource Inventory.* Vancouver: UBC Press. 754 pp.

Vertrees, J.D. 1978. *Japanese Maples.* Portland, OR: Timber Press. 178 pp.

Whitelaw, E. & C. Sihoe. 1983. 'Vancouver's Heritage Tree Inventory – A Collection of Great Trees in the City.' Sponsored by the British Columbia Society of Landscape Architects. Unpubl. ms. 167 pp.